Effect of wheat bran on the expansion and mechanical properties of extruded plasticized starchy foams

zur Erlangung des akademischen Grades eines
DOKTORS DER INGENIEURWISSENSCHAFTEN (Dr.-Ing.)

der Fakultät für Chemieingenieurwesen und Verfahrenstechnik des
Karlsruher Institut für Technologie (KIT)

genehmigte

Dissertation

von

Frédéric Robin, M.Sc.
aus Villeurbanne, Frankreich

Referent: Prof. Dr.-Ing. H.P. Schuchmann (Karlsruher Institut für Technologie)
Korreferent: Prof. Dr-Ing. habil. S. Palzer (Nestec S.A.)
Tag des Kolloquiums: 01.09.2011

Bibliographic information published by the Deutsche Nationalbibliothek

The Deutsche Nationalbibliothek lists this publication in the Deutsche Nationalbibliografie; detailed bibliographic data are available in the Internet at http://dnb.d-nb.de .

ISBN 978-3-8325-3078-5

Logos Verlag Berlin GmbH
Comeniushof, Gubener Str. 47,
D-10243 Berlin
Tel.: +49 (0)30 42 85 10 90
Fax: +49 (0)30 42 85 10 92
INTERNET: http://www.logos-verlag.de

Berichte aus der Verfahrenstechnik

Frédéric Robin

Effect of wheat bran on the expansion and mechanical properties of extruded plasticized starchy foams

Une illumination soudaine semble parfois faire bifurquer une destinée. Mais l'illumination n'est que la vision soudaine, par l'Esprit, d'une route lentement préparée.
[Antoine de Saint-Exupéry]

A ma Famille et mes amis: passés, présents et futurs…

ACKNOWLEDGMENTS

First of all, I would like to thank Prof. Peter van Bladeren, Head of the Nestlé Research Center, for giving me the opportunity to perform this work in parallel to my industrial employment.

I would also express my sincere gratitude to my supervisors: Prof. Dr.-Ing. habil. Stefan Palzer and Prof. Dr.-Ing. Heike P. Schuchmann. I would like to thank them for their trust and belief in my capacities to achieve this work, guidance and support, as well as for their high scientific level.

A special thank goes to Dr. Monica Fischer, Dr. Simon Livings, Dr. Undine Lehmann and Dr. Stefan Kaufmann for encouraging and motivating me to do a PhD as well as offering me the liberty to do so.

My further thank goes to the persons who participated to the bench experiments and data analysis of this work: Adeline Boire, Steffen Dattinger, Delphine Curti, Dr. Cédric Dubois, Dr. Jan Engmann, Dr. Alessandro Gianfrancesco, Chloé Merlin, Dr. Nicolas Pineau, Dr. Laurent Forny, Christine Théoduloz and Dr. Gilles Vuataz. A special thank goes to Nicolas Bovet for all the long hours we spent in the pilot plant running the extruder and the on-line rheometer and to Alain Fracheboud for his "there is always a solution" way of thinking.

I strongly thank my colleagues from the Nutrient Technology group who supported me from the very beginning of my start at the Nestlé Research Center. They have always been available to give me support on analytical measurements or answer my questions. I specially thank Dr. Olivier Roger, Dr. Robert Redgwell and Dr. Christophe Schmitt, my office mates, for our discussions about dietary fibers and for standing to my endless questions and thoughts about science and… Life. Big thank to Dr. Hugo Gloria-Hernandez and Dr. Sathaporn Srishuwong for their support and advice on DSC measurements and starch.

I thank as well my fellow PhD students at the Karlsruhe Institute of Technology and especially Azad Emin, Mario Horvat and Mario Hirth for their warm welcome, scientific discussions about extrusion and nice atmosphere they created during my stays in Karlsruhe.

Last, but not the least, I would like to thank my Family and my friends for their patience and support especially when writing during evenings and weekends.

ZUSAMMENFASSUNG

Die Nachfrage der Verbraucher nach gesünderen Lebensmitteln ist weiterhin hoch. Weizenkleie, ein leicht zugängliches und kostengünstiges Nebenprodukt, enthält einen hohen Anteil an Ballaststoffen, welche ein ernährungs- und gesundheitsbezogenen Mehrwert aufweisen. Dennoch ist ihre Verwertung in extrudierten Nahrungsmitteln aufgrund geringerer Expansionsvolumen und negativer Beeinflussung der strukturellen Eigenschaften eingeschränkt. Um die Verwendung von Weizenkleie zu verbessern, ist das Verständins und die Kontrolle ihrer Auswirkungen auf die texturbeeinflussenden Parameter von extrudierten Produkten ausschlaggebend.

Das Ziel dieser Arbeit war es, den Einfluss von Kleie auf die für die Bildung von extrudierten stärkehaltigen Schäumen verantwortlichen Parameter und deren mechanische Eigenschaften zu untersuchen. Zu diesem Zweck wurde Weizenmehl mit Weizenkleie angereichert. Die verschiedenen Rezepte wurden, einem Versuchsplan folgend, bei unterschiedlichen Schneckendrehzahlen, Gehäusetemperaturen und Anfangswassergehalten extrudiert. Die Wirkung von Kleie auf die die Expansion beeinflussenden Parameter wurde mittels Röntgen-Tomographie, on-line Schmelzrheologie und dynamischer Kalorimetrie untersucht. Die Glasübergangstemperaturen und physikalisch-chemischen Eigenschaften der Stärke wurden bestimmt. Die mechanischen Eigenschaften wurden mittels Drei-Punkt-Biegeversuche analysiert.

Die Erhöhung der Kleiekonzentration in Weizenmehl begünstigt die Längenausdehnung zum Nachteil der volumetrischen und radialen Ausdehnung. Diese Längenausdehnung ist mit der Erzeugung feiner Strukturen mit einer höheren Dichte von kleinen Zellen verbunden. Bei ähnlichen relativen Dichten führen diese feineren Strukturen zu Produkten mit höherer Festigkeit. Die Zunahme der Zelldichte wird durch einen höheren Keimbildungsgrad an der Düse hervorgerufen. Eine Erhöhung der Scherviskosität tritt nur bei der höchsten Kleiekonzentration auf, was auf eine Schwellen-Kleiekonzentration hindeutet. Dieser Anstieg der Scherviskosität könnte die Expansion behindern. Bei Erhöhung der Kleiekonzentration sinkt die Glasübergangstemperatur der Stärke, da der amorphen Stärke mehr freies Wasser zur Verfügung steht. Diese Abnahme der Glasübergangstemperatur könnte zu einer geringeren Viskosität der Stärkephase führen und der erhöhten Matrix-Scherviskosität im Zusammenhang mit den Kleie-Eigenschaften entgegenwirken. Durch den Zusatz von Kleie ändert sich der Eingangsdruckverlust an der Düse, gemessen durch verschiedene Lochblenden am Düsenauslass, deutlich. Dies deutet auf eine Verringerung der elastischen Eigenschaften hin. Bei einer Erhöhung der Kleiekonzentration scheint Stärke in einem höheren Ausmaß umgewandelt zu werden. Die erhöhte Stärke-Transformation kann auch zu Veränderungen in der Schmelzrheologie beitragen. Eine erhöhte Oberflächenporosität der Proben ist bei der Erhöhung der Weizenkleie-Konzentration beobachtet worden. Diese höhere Porosität der Oberfläche ist durch ein Aufbruch der Blasen während des Wachstums zu erklären. Eine reduzierte

Expansion der kleiehaltigen Proben ist die Folge. Dieses Phänomen wird durch feinere Blasenwände begünstigt und kann durch die begrenzte Haftung zwischen den Kleiepartikeln und der kontinuierlichen Stärkephase erklärt werden.

Diese Arbeit zeigt, dass die Beimischung von Kleie zu extrudierten stärkebasierten Produkten zu einer Verringerung der Expansionsvolumina und Strukturen mit höherer mechanischer Festigkeit führt. Kleiepartikel wirken als Füllstoffe, welche nur geringfügig durch Stärke benetzt werden. Dies veränderte die stärkehaltige Schmelzrheologie und destabilisiert die Blasenwände während des Wachstums. Die zukünftige Forschung sollte sich daher auf die Veränderungen der rheologischen Eigenschaften von Weizenkleie und/oder ihrer physikalisch-chemischen Verträglichkeit mit Stärke vor oder während der Extrusion konzentrieren.

ABSTRACT

Consumers demand for healthier food products remains high. Wheat bran, a readily available and low cost by-product, contains a high amount of dietary fibers reported to deliver nutritional and health benefits. Nevertheless, its valorization in extruded food products is limited due to its small expansion volumes and negative effect on textural properties. In order to enhance the use of wheat bran, it is important to understand and control its effect on the parameters driving textural properties of extruded products.

The objective of this work was to investigate the effect of bran on the parameters driving the formation of extruded starchy foams and on their mechanical properties. For this purpose, refined wheat flour was enriched with wheat bran. The recipes were extruded at varying conditions of screw speed, barrel temperature and water content in the feed according to an experimental plan. The effect of bran on the drivers of expansion was assessed using micro-computed X-ray tomography, on-line melt rheology and dynamical scanning calorimetry. The glass transition temperatures and physicochemical properties of the starch were determined and the mechanical properties were assessed using three-point bending tests.

Increasing the bran concentration in wheat flour favors longitudinal expansion at the expense of volumetric and sectional expansions. This longitudinal expansion is associated with the generation of finer structures with a higher density of small cells. At similar relative densities, these finer structures lead to products with a higher mechanical strength. The increase in cell density is induced by a higher degree of nucleation at the die. Increase in shear viscosity occurs only at the highest bran concentration, indicating a threshold bran concentration. This increase in shear viscosity may hinder expansion. The glass transition temperature of starch is lowered when increasing the bran concentration. This is due to more free water available for the amorphous starch. This decrease in glass transition temperature could lead to a lower viscosity of the starch phase and counteract the increased matrix shear viscosity linked to the bran properties. The entrance pressure drop at the die, measured by the die orifice method, is significantly modified by the addition of bran. This may indicate reduced elastic properties. Starch appears to be transformed to a higher extent when increasing the bran concentration. This higher degree of starch transformation may also participate to changes in the melt rheology. An increased surface porosity of the samples is observed when increasing wheat bran concentration. This higher porosity of the surface can be explained by a rupture of the bubbles during growth. Such a rupture contributes to the reduced expansion of bran-containing samples. This phenomenon is favored by finer bubbles walls and can be explained by the limited adhesion between the bran particles and the continuous starch phase.

Incorporation of bran in extruded starchy-based products leads to a reduction in expansion volumes. Bran particles act as fillers that are not well wetted by starch. This modifies rheology of the melt and destabilizes the bubble walls during growth. Future research should focus on solutions to modify the rheological properties of wheat bran and/or its physicochemical compatibility with starch prior or during extrusion.

The results discussed in this manuscript are based on the work published or submitted in:

Robin, F., Engmann, J., Tomasi, D., Breton, O., Parker, R., Schuchmann, H. P., Palzer, S. (2010). Adjustable twin-slit rheometer for shear viscosity measurement of extruded complex starchy melts. *Chemical Engineering & Technology*, 33 (10), 1672-1678

Robin, F., Théoduloz, C., Gianfrancesco, A., Pineau, N., Schuchmann, H. P, Palzer, S. (2011). Starch transformation in bran-enriched extruded wheat flour. *Carbohydrate Polymers*, 85, 65-74

Robin, F., Bovet, N., Pineau, N., Schuchmann, H. P, Palzer, S. (2011) On-Line Shear Viscosity Measurement of Starchy Melts Enriched in Wheat Bran. *Journal of Food Science*, 76 (5), E405-E412

Robin, F., Dubois, C., Pineau, N., Schuchmann, H. P, Palzer, S. (2011). Expansion mechanism of extruded foams supplemented with wheat bran. *Journal of Food Engineering,* 107, 80-89

Robin, F., Dubois, C., Curti, D., Schuchmann, H. P, Palzer, S. (2011). Effect of wheat bran on the mechanical properties of extruded starchy foams. *Food Research International*, 44, 2880-2888

Robin, F., Dubois, C., Schuchmann, H. P, Palzer, S. (2011). Supplementation of extruded foams with wheat bran: Effect on textural properties. Processing ICEF11, 22-26, 2011, Proceedia, 1, 505-512

Robin, F., Dattinger, S., Boire, A., Horvat, M., Schuchmann, H. P, Palzer, S. (2012). Elastic properties of extruded starchy melts containing wheat bran using on-line rheology and dynamic mechanical thermal analysis. *Journal of Food Engineering*, 109 (3), 414-423

and based on the results presented in:

Robin, F. Dubois, Curti, D., Kaufmann, S.F.M., Schuchmann, H. P., Palzer, S. Expansion and material properties of extruded foams containing wheat bran, IUFost 2011 congress, South Africa, poster session, August 23, 2010

Robin, F., Dubois, C., Schuchmann, H. P., Palzer, S. Expansion properties of extruded foams containing wheat bran. Oral presentation, Jahrestreffen des Fachausschusses Lebensmittelverfahrenstechnik 2011, Vlaardingen, Netherlands, March 22-24, 2011

Robin, F., Dubois, C., Schuchmann, H. P., Palzer, S. Supplementation of extruded foams with wheat bran: Effect on textural properties. Oral presentation, International Congress on Engineering and Food, ICEF11, Athens, May, 22-26, 2011

Robin, F., Schuchmann, H. P., Palzer, S. Effect of wheat bran on the expansion and mechanical properties of extruded plasticized starchy foams. Oral presentation, European Congress of Chemical Engineering 2011, Berlin, 25-29 September, 2011

KEY WORDS

Extrusion, cereals, wheat bran, expansion, mechanical properties, melt rheology

TABLE OF CONTENTS

ABBREVIATIONS & SYMBOLS

ABBREVIATIONS

		[SI Units]
d.m.	Dry matter	[-]
IDF	Insoluble dietary fiber content	[%]
LEI	Longitudinal Expansion Index	[-]
MCS	Mean Cell Size	[m]
MCWT	Mean Cell Wall Thickness	[m]
MC	Moisture content	[%]
MC_e	Moisture content of the extrudate	[%]
MC_m	Moisture content of the melt	[%]
SEI	Sectional Expansion Index	[-]
SEI_{max}	Maximum Sectional Expansion Index at the die	[-]
SR	Shrinkage Ratio	[%]
SWS	Estimated Starch Water Solubility	[%]
SDF	Soluble dietary fiber content	[%]
SME	Specific Mechanical Energy	[kJ kg^{-1}]
TDF	Total dietary fiber content	[%]
VEI	Volumetric Expansion Index	[-]
VOI	Volume of Interest	[m^3]
WAI	Water Absorption Index	[-]
WSI	Water Solubility Index	[%]
Y	Expansion properties	
DSC	Dynamic Scanning Calorimeter	
LB	Low bran concentration	
HB	High bran concentration	
G&T	Gordon & Taylor	
LSD	Leas significant difference	
RF	Refined flour	
RVA	Rapid visco-analysis	
WLF	Williams-Landel-Ferry Model	
XRT	X-ray tomography	

SYMBOLS **[SI Unit]**

A	Cell area	[m^2]
a_w	Water activity	[-]
B_{WLF}	WLF constant	[K]
C_{WLF}	WLF constant	[-]
C_{GAB}	GAB Constant	[-]
d	Maximum displacement at rupture	[m]
D	Relative density	[-]
D_d	Die diameter	[m]
D_e	Extruder screw diameter	[m]
D_{max}	Maximum diameter of the expandate at the die exit	[m]
$D_{[4, 3]}$	Average volume weighted particle size	[m]
ΔP	Pressure difference between pressure sensors	[Pa]
ΔP_{end}	Capillary end pressure drop	[Pa]
ΔP_{cap}	Pressure drop over the capillary length	[Pa]
ΔP_{Ent}	Pressure drop at the capillary entrance	[Pa]
ΔP_{Exit}	Pressure drop at the capillary exit	[Pa]
ΔE	Activation energy	[J mol^{-1}]
E^*	Elastic modulus of expanded foams	[Pa]
E_s	Elastic modulus of the unexpanded material	[Pa]
F	Maximum force at rupture	[N]
H	Rheometer slit height	[m]
I_{co}	Connectivity index	[%]
J_{hom}	Nucleation rate for homogenous nucleation	[s^{-1}]
J_{het}	Nucleation rate for heterogeneous nucleation	[s^{-1}]
k_B	Boltzmann's constant	[mol^{-1}]
$k_{G\&T}$	G&T constant	[-]
K_{GAB}	GAB constant	[-]
K	Consistency factor	[Pa s]
K_0	Constant viscosity models	[Pa s]
l	Cell edge length	[m]
L	Capillary/Slit length	[m]
L_d	Die length	[m]
L_e	Extruder barrel length	[m]
L_p	Piston length	[m]
L_b	Distance between supports for the bars	[m]
L_s	Distance between supports for the extruded samples	[m]
L_{se}	Specific length of extruded sample	[m kg^{-1}]
M	Motor torque	[N m]
M	Molecular weight	[kg mol^{-1}]
MC	Moisture content	[%]

m_{total}	Total mass flow rate	[kg h^{-1}]
M_{unload}	Motor torque without load	[N m]
M_d	Water content	[%]
N	Screw speed	[rpm]
N_c	Cell density	[m^{-3}]
N_d	Atom density	[kg m^{-3}]
N_1	First normal stress difference	[Pa]
n	Power law index	[-]
n_{act}	Actual screw speed	[rpm]
n_{max}	Maximum screw speed	[rpm]
n_B	Bagley correction	[-]
P	Sample porosity	[%]
P_0	Pressure at the rheometer entrance	[Pa]
P_d	Pressure in the front plate with the attached die	[Pa]
P_{max}	Maximum motor power	[kW]
P_V	Pressure of the vapor phase	[Pa]
P_L	Pressure of the liquid phase	[Pa]
P_r	Pressure in the front plate with the attached rheometer	[Pa]
P_i	Pressure from slits pressure sensors	[Pa]
Q	Volumetric flow rate in left or right slit	[kg m^3]
$Q_{1/2}$	Volumetric flow rate in left or right slit	[kg m^3]
Q_m	Volumetric flow rate measured in a capillary	[m^3 s^{-1}]
Q_L	Volumetric flow rate in the left slit	[m^3 s^{-1}]
Q_R	Volumetric flow rate in the right slit	[m^3 s^{-1}]
Q_T	Total volumetric flow rate	[m^3 s^{-1}]
Q_{ws}	Volumetric flow rate without slip	[m^3 s^{-1}]
R	Gas constant	[J kg^{-1} mol^{-1}]
R	Radius of bubbles/capillary rheometer	[m]
R	Shape-anisotropy ratio	[-]
r_e	Extruded tube radius	[m]
t	Cell edge thickness	[m]
T	Temperature	[K]
T_d	Temperature in the extruder front plate with die attached	[K]
T_g	Glass transition temperature	[K]
$T_{g.\ onset}$	Temperature at the onset glass transition	[K]
$T_{g,m}$	Glass transition of fully dried material	[K]
$T_{g,w}$	Glass transition of pure water	[K]
T_m	Melting temperature	[K]
T_{pk}	Melting temperature at the endothermic peak	[K]
T_i	Pressure from slits temperature probes	[K]
x_r	Distance between two temperature probes	[m]
y_r	Distance between two temperature probes	[m]
V_1	Molar volume of water	[mol m^{-3}]

V_2	Molar volume of polymer	[mol m^{-3}]
u_s	Effective slit velocity	[m s^{-1}]
$W(a_w)$	Water content at constant water activity	[%]
W_m	Monolayer water content	[%]
W	Rheometer slit width	[m]
W_b	Thermomolded bar width	[m]
x_r	Distance between two pressure sensors	[m]
x, y, z	Axis direction	[-]

GREEK SYMBOLS		[SI Unit]
β	Effective slit coefficient	[m^3 Pa^{-1} s^{-1}]
β_c	Corrected effective slit coefficient	[m^2 Pa^{-1} s^{-1}]
ΔP	Pressure drop	[Pa]
χ_{12}	Flory-Huggins interaction polymer-diluent parameter	[-]
ε	Porosity	[%]
$\dot{\varepsilon}$	Extensional rate	[s^{-1}]
ε^*	Strain of expanded samples	[%]
ε_s	Strain of unexpanded material	[%]
ϕ	Contact angle of a liquid with a surface	[°]
ϕ	Fraction of solid contained in cell edges	[-]
ϕ_{max}	Maximum packing factor	[-]
$\dot{\gamma}_a$	Average shear rate	[s^{-1}]
$\dot{\gamma}$	Corrected shear rate	[s^{-1}]
$\dot{\gamma}_a$	Apparent shear rate	[s^{-1}]
$\bar{\gamma}_d$	Total shear train at the single-hole die	[s^{-1}]
$\bar{\gamma}_{ms}$	Total shear train in the metering session	[s^{-1}]
η	Viscosity	[Pa]
η_{RVA}	RVA peak viscosity	[Pa]
$\eta(\dot{\gamma})$	Shear viscosity	[Pa s]
η_{pa}	Apparent extensional viscosity	[Pa s]
η_a	Apparent shear viscosity	[Pa s]
η_0	Reference viscosity	[Pa s]
$(\mu_a)_d$	Apparent viscosity at the die	[Pa s]
v_1	Volume fraction of water	[-]
v_2	Volume fraction of polymer	[-]
ρ^*	Bulk density of the foam	[kg m^{-3}]

ρ_e	Density of the extrudate	[kg m^{-3}]
ρ_m	Density of the melt	[kg m^{-3}]
ρ_s	Bulk density of the cell walls	[kg m^{-3}]
σ	Interfacial or surface tension	[N m^{-1}]
σ	Shear stress	[Pa]
σ^*_{cr}	Stress at rupture of expanded foam	[Pa]
σ_s	Stress at rupture of unexpanded material	[Pa]
σ^*	Stress at rupture of expanded samples	[Pa]
σ_w	Shear stress at the capillary wall	[Pa]

CHAPTER I. INTRODUCTION

Cooking-extrusion is one of the main technologies of the Food Industry to process cereals. It is a cost efficient process that offers continuous production, flexibility and opportunities to broaden the range of textures and shapes of processed dough. First inspired from the synthetic polymer industry, it has been used for more than half a century to produce puffed food products like breakfast cereals or savory snacks (Eastman, 2001). Its added value to product properties is to transform a non edible, difficult to digest, dense material into a light aerated crispy and easy-to-digest product.

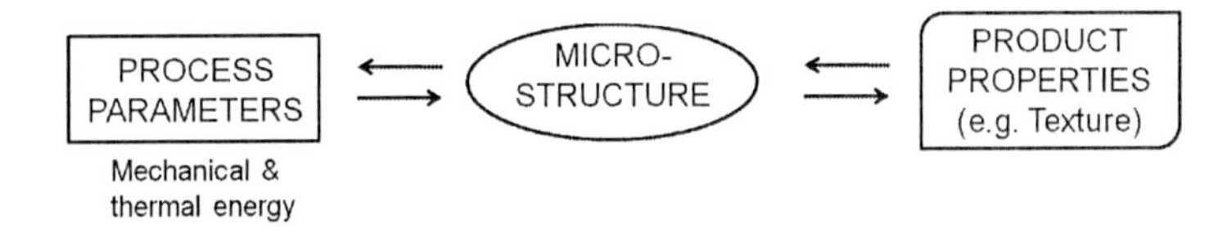

Fig. 1.1: Interacting effects of the ingredients and extrusion conditions on microstructure of puffed extruded starch-based products and their properties

The principle of cooking-extrusion (with direct expansion) is based on the transformation of cereals under pressure. A semi-crystalline starchy material is transformed into a molten mass, easy to digest and called "melt". At the outlet of the extruder, the pressure is relieved, leading to a sudden expansion of the melt. Cooking-extrusion is a complex process driven by interdependent parameters (see Fig. 1.1). The expansion phenomenon of the molten material, occurring at the die exit and leading to the formation of aerated foams, has been described as a multistep mechanism (Moraru and Kokini, 2003). The properties of puffed foods are primarily governed by the physicochemical properties of the ingredients, especially of starch, and their transformations during extrusion. They then drive the rheological properties of the melt and its expansion at the die outlet. The bulk expansion properties and cellular microstructure characteristics resulting from the expansion are then responsible for the final textural attributes of the product (see Fig. 1.1). Numerous studies have been carried out to better understand and control the parameters driving expansion and texture of extruded cereals. Nevertheless, this is still an area of open questions. Especially processes taking place in the extruder itself and influence of non-starchy ingredients on expansion are still unknown.

In the last two decades, overweight population and obesity issues have grown fast. They have become major concerns for the public health authorities and governments. The consumption habits have drastically changed over the years and may contribute to this result. Nevertheless, nutritionally unbalanced foods, favoring high fat and carbohydrate-containing products, have also been pinpointed by the authorities. Large communication campaigns have raised consumers' awareness of the composition of their food and advantages of a healthier lifestyle and consumption habits for their health. Within this frame, the Food Industry is looking for solutions to respond to the consumer demand for healthier foods. Increasing the dietary fiber content in foods to reduce their calorie content and improve their nutritional value is one of the main strategic directions. In addition to their low calorie content and regulation of digestion, dietary fibers were also shown to have secondary health benefits. These benefits include cholesterol lowering effects and benefits linked to weight management (Marlett et al, 2002). However, the daily consumption of dietary fibers is still much lower than the daily recommended amount (Marlett et al, 2002). This is linked to the unclear and inconsistent communication over health benefits of dietary fibers, product availability and/or unpleasant product texture, taste and appearance (Hamaker, 2008). Although consumers are looking for healthier food, they are not ready to compromise on sensorial attributes (taste and texture) for health.

Increasing dietary fiber content in extruded direct expanded cereals is challenging. This is due to the reduced expansion properties of fibers compared to starch. Nevertheless, all dietary fibers do not show the same effect on expansion and final product properties. Soluble fibers such as inulin deliver a higher expansion than insoluble fibers such as cereal fibers (Blake, 2006). Still, cereals fibers are less expensive and largely available. Wheat bran is a by-product of the wheat grain refining process and contains a high amount of dietary fibers. It is a readily available and low-cost ingredient that is mostly used for animal feeding. Its effect on expansion volumes, leading to less preferred textures by consumers, is well known and has been investigated in numerous studies (e.g. Lue et al., 1990; Moore et al., 1990; Wang et al., 1993, Yanniotis et al., 2007; Brennan, 2008). However, only few studies focused on its impact on the expansion mechanism of extruded foods. Identifying, quantifying and explaining the effects of bran on the parameters driving expansion and texture of the final product are necessary to provide solutions to the Industry for an optimal design of bran-containing extruded products.

This work was focused on the cooking-extrusion of wheat flour enriched in wheat bran. The effect of bran addition on physicochemical changes of starch during extrusion, the melts rheological and expansion properties as well as the mechanical properties of the resulting foams were investigated. Accordingly, wheat flour was supplemented with wheat bran to a maximum of 24.4 % dietary fibers). This fiber concentration was chosen to obtain 3 g (w.b.) of dietary fibers in extruded breakfast cereals with a serving size of 30 g and at a total cereal content of 50-60 % (w/w). Reaching 3 g of dietary fibers in the product enable to display content claims on the product packaging. The bran containing samples and the refined wheat flour samples (used as reference) were extruded at different conditions of barrel temperature, screw speed and water content in the feed. Trials were performed according to an experimental design. The effect of the bran particle size on expansion was also investigated. The experimental plan enabled changing the conditions in a systematic way and identifying the main effect of process parameters and bran concentration on product properties as well as their interactions. The extrusion conditions were chosen broad to generate a wide range of expansion properties and cellular structures. In this study, the effect of wheat bran on the expansion of extruded plasticized starch and on the mechanical properties of the resulting foams was evaluated using a mechanistic and material science approach. Scientific knowledge of synthetic polymer extrusion was also used to explain the influence of bran on the expansion mechanism. The study was performed applying analytical tools new to such complex product compositions. These tools include X-ray tomography and on-line twin-slit viscometry.

In chapter II, the State of the Art in the field of cooking-extrusion of cereals focusing on the different mechanistic steps of expansion is presented. The role of rheological properties in the expansion mechanism is highlighted. The literature dealing with addition of fibers in extruded cereals and scientific gaps in this field are also summarized. The material and methods are described in Chapter III. The physicochemical changes of starch induced by the addition of wheat bran and their dependency on process conditions are presented in Chapter IV. In Chapter V, the use and set-up of a twin-slit rheometer to investigate the effect of wheat bran on the shear viscosity of the melt are reported. The effect of bran on the entrance pressure drop of extruded wheat flour using an on-line slit rheometer is discussed. The relationship between the rheological properties and the expansion properties is discussed in Chapter VI. In Chapter VII, it is reported the effect of wheat bran on mechanical properties of the foams generated by extrusion. Finally, in Chapter VIII the main findings of this work are summarized leading to open questions for future researches.

CHAPTER II. STATE OF THE ART

II.1. Introduction

Incorporating wheat fibers in extruded products enables to food companies to communicate on nutritional and health benefits to consumers. Nevertheless definition of fibers, allowed content claims and communication over health benefits obey rules defined by regulatory bodies and country laws. The definition of fibers in extruded foods is of special importance as it is based on the physicochemical properties of fibers that may be modified during extrusion. In order to evaluate the effect of bran on the expansion properties of starchy foams, the parameters driving these properties need to be identified. Among these parameters, the rheological properties of the melt are essential. On-line rheometers enable to assess the rheology of the melt in the conditions of extrusion. The textural properties of an extruded product are the result of the expansion properties. These are driven by several physicochemical parameters including shape, dimensions, porosity (cell size, shape and density) and cell wall thicknesses of the product. These parameters may be altered by the addition of wheat bran.

This Chapter reviews the current State of the Art concerning the parameters driving the expansion and mechanical properties of solid foams. The analytical tools used to assess the parameters driving expansion and mechanical properties are also discussed. Based on this review of the current scientific literature, the scientific gaps in the field of extruded foams containing wheat bran are highlighted.

II.2. Dietary fiber definition, sources, health benefits and regulatory aspects

The definition of dietary fibers has been evolving through years until Codex Alimentarius finally adopted a definition in its 2009 meeting. The definition is the following (Phillips, 2011):

"Dietary fibers means carbohydrate polymers with ten or more monomeric units which are not hydrolyzed by the endogenous enzymes in the small intestine of humans and belong to the following categories:

- Edible carbohydrate polymers naturally occurring in the food as consumed
- Carbohydrate polymers, which have been obtained from food raw material by physiological, enzymic or chemical means and which have been shown to have a physiological effect of benefit to health as demonstrated by generally accepted scientific evidence to competent authorities
- Synthetic carbohydrate polymers which have been shown to have a physiological effect of benefit to health as demonstrated by generally accepted scientific evidence to competent authorities

Dietary fibers found in the human diet are differentiated according to their water solubility. Insoluble fibers represent about 75 % of the dietary fiber found in foods. They consist mainly of cell wall components such as cellulose, lignin, and hemicelluloses present in wheat, most grain products and vegetables. Insoluble fibers shorten bowel transit time, increase fecal bulk and render feces softer. Soluble fibers consist of noncellulosic polysaccharides such as pectin, gum and mucilage found in fruits, oats, barley and legumes. Soluble fibers delay gastric emptying, slow glucose absorption, enhance immune function and lower serum cholesterol levels. They are to a large degree fermented in the colon into short-chain fatty acids, which may inhibit hepatic cholesterol synthesis (Dreher, 2001).

The daily recommended dietary fiber intake for adults is in the range of 20 to 35 g or 10 to 13 g of dietary fiber per 1000 kcal, depending on the country and gender. Nevertheless the average daily intake is less than the recommended one (Marlett et al., 2002). This is the case for instance in the United States where the daily intake for adults is in the range of 14 to 15 g of dietary fiber (Marlett et al., 2002). Communication of the nutritional benefits to consumers through quantitative claims depends on the country. The Codex recommends that any product claiming to be a 'source' of fiber should contain 3 g of fiber per 100 g of serving or 1.5 g of fiber per 100 kcal of serving or 10 % of daily reference value per serving. To claim that a food is high in fiber, the product must contain at least 6 g of fiber per 100 g of serving or 6 g of fiber per 100 kcal of serving or 20 % of daily reference value per serving (Anonymous, Codex Guidelines, 2009).

II.3. Wheat bran and fibers

Wheat is the main cereal used in human nutrition. It is estimated that approximately one-third of the world's population depends upon wheat for their nourishment (Hamaker, 2008). The Food and Agriculture Organization (FAO) of the United Nations estimated a global production of wheat for 2009/2010 of about 682 million of tons (FAO, 2010).

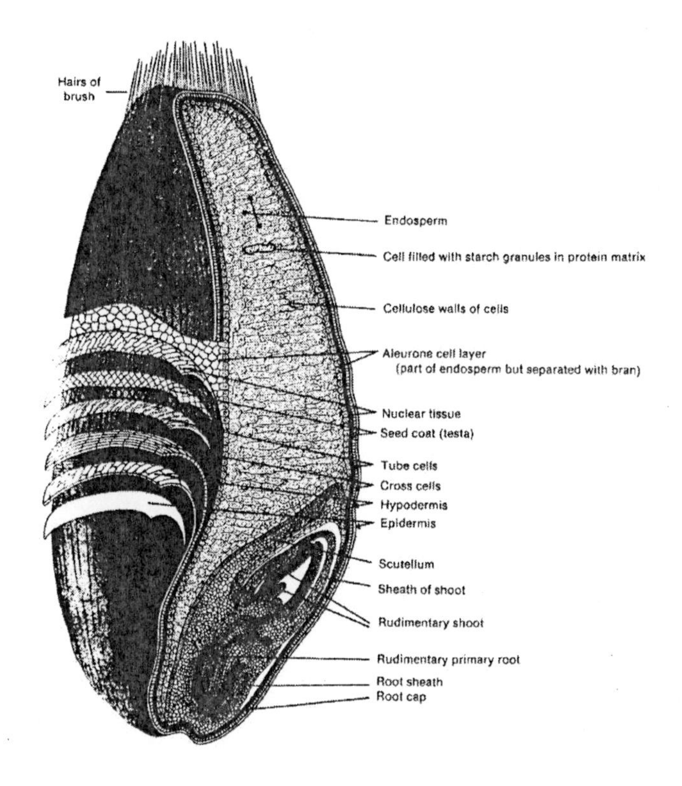

Fig. 2.1: Longitudinal section of the wheat kernel (adapted from Delcour and Hoseney, 2010)

The wheat grain consists of three physical parts: the endosperm, the bran and the germ (Fig. 2.1). The endosperm represents the main part of the grain's volume and weight. It mainly contains starch granules surrounded by a protein matrix. The bran part consists of the outer coasts of the grain and contains mainly fibers (Cornell and Hoveling, 1998). The bran accounts for approximately 12-15 % of the grain (Cho and Clark, 2001). The germ is rich in vitamins, proteins and lipids and accounts for about 3-5 % of the weight of the grain (Cornell and Hoveling, 1998). The main milling technique used to separate the different parts of the grain is roller milling. During roller milling, the bran is flattened and disintegrated and the endosperm material is reduced in size (Cornell and Hoveling, 1998). The commercial bran fraction obtained from the mill is not homogeneous with respect to tissues and biochemical compositions. Some of the wheat germ and a small amount of endosperm remains because of the difficulty of making a sharp separation. Wheat bran contains in general about 45-50 % of dietary fiber. Ralet et al. (1990) reported that about

80 % of the total dietary fibers in wheat bran are insoluble while 20 % are soluble. Wheat bran contains approximately the same amount of proteins than in the whole wheat flour, but it is richer in ash, lipid, and particularly in pentosans (Cornell and Hoveling, 1998).

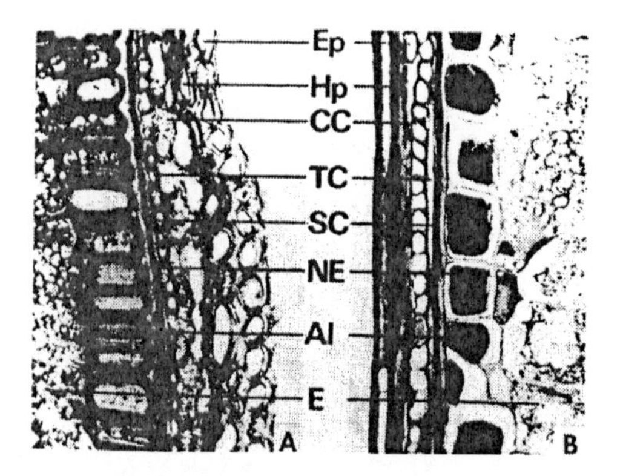

Fig. 2.2: **Pictures of the outer layers of the wheat grain sectioned transversally (A) and longitudinally (B): Ep = epidermis, Hp = hypodermis, CC = cross cell, TC = tube cell, SC = seed coat, NE = nucellar epidermis, Al = aleurone layer, E = starchy endosperm (adapted from Pomeranz, 1988)**

The outer coasts of bran successively, from the outer to the inner surface, comprise the outer pericarp (beeswing bran), the inner pericarp (comprising cross cells and tube cells), seed coat (testa), hyaline layer (nucellar epidermis) and the aleurone layer (Fig. 2.2). The aleurone layer is the thickest (up to 65 μm), the outer pericarp is of intermediate thickness (15-30 μm) and the seed coat is the thinnest layer (5-8 μm) (Barron, 2007). Cellulose, usually associated with glucommannans, soluble (1→3, 1→4)-*β*-glucan and arabinoxylans are the main non-starch polysaccharides present in the lignified bran layers (Pomeranz, 1988). The precise composition of the bran layers are not exactly known due to the difficult in separating the different layers. The aleurone layer is composed of ~ 2 % cellulose (w/w), ~ 7 % (w/w) glucomannan, ~ 29 % (1→3, 1→4)-*β*-glucan and ~ 65 % arabinoxylan (Rodhes et al., 2002). Aleurone layer walls are heavily esterified with hydroxycinnamic acids (1.8 % w/w), which are associated with the wall's arabinoxylans.

II.4. Cooking-extrusion of cereal products

Cereal grains need to be processed to be edible. Extrusion-cooking is used to transform cereal grains and derivates (e.g. flour, starch) in an aerated edible product. In extrusion-cooking, a semi-crystalline starchy powder is transformed in a continuous viscous molten mass by the application of a combination of mechanical and thermal energy. At the exit of the extruder, restriction of flow leads to pressure build-up. At the die exit, the pressure is relieved and the superheated water flashes off, leading to an expanded foamy product called "extrudate".

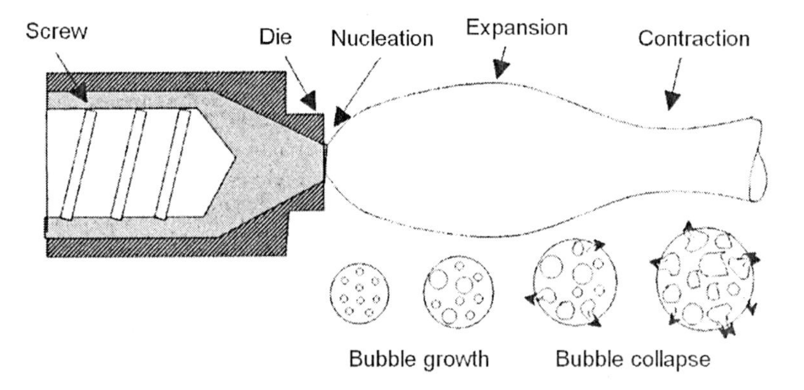

Fig. 2.3: Schematic illustration of extrudate expansion (adapted from Kokini et al., 1992)

Moraru and Kokini (2003) described the expansion mechanisms of extruded starchy products as a multistep process consisting of 5 major steps: starch transformations, nucleation of bubbles, extrudate swell, bubble growth and bubble collapse (Fig. 2.3). These different mechanistic steps are affected by the addition of wheat bran. This may explain the differences in expansion properties reported in literature (see § II.5). The different steps of expansion are detailed in the following subchapters.

II.4.1. Starch transformations during extrusion

Starch is the main macromolecule in cereals. It is a semi-crystalline granule composed of two polymers: amylose and amylopectin. Amylose is essentially composed of linear molecules of α(1-4)-linked D-glucopyranose with a polymerization degree in the order of 1500-6000 (Zobel, 1988). Amylopectin is characterized by branch chains of α(1-4)-linked D-glucose chains with different length and connected by α(1-6)-linkages. Its degree of polymerization ranges between 3.10^5 and 3.10^6 (Zobel, 1988). During extrusion of starch-based products, starch undergoes supramolecular and molecular changes depending on the process conditions. Several analytical tools are available to evaluate the physicochemical changes of starch. Among them water solubility index (WSI) and water absorption index (WAI) tests are used the most. The water solubility index represents the content of material that is soluble in cold water. The cold water binding capacity of the material is obtained measuring the water absorption index. Both indices allow estimating the degree of transformation of starch and other ingredients due to interactions with water.

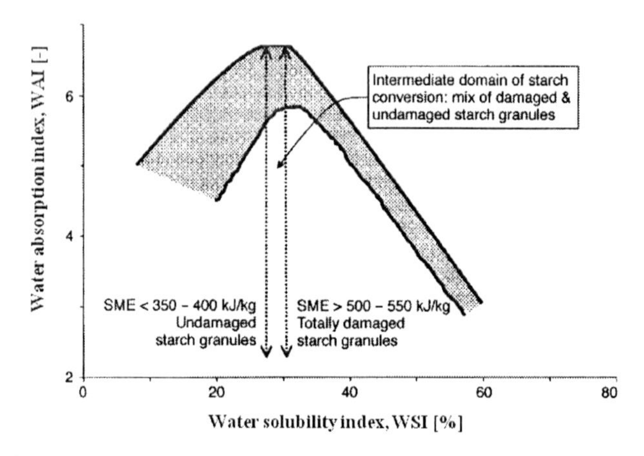

Fig. 2.4: Relationship between water absorption index (WAI) and water solubility index (WSI) of maize grits according to the specific mechanical energy (SME) (adapted from Smith, 1992)

Smith et al. (1992) demonstrated that a general relationship between water absorption index (WAI) and water solubility index (WSI) exists for extruded starchy material. The correlation between these two parameters can be seen in Fig. 2.4. In extrusion process, this relationship depends on the specific mechanical energy (SME). The specific mechanical energy (SME) is the energy going to the extrusion system per unit mass in the form of work. It is primarily calculated from the engine torque and screw speed. This energy can be related to several product properties; e.g. the melt viscosity (e.g. van Lengerich and Larson, 2000; Schuchmann and Danner, 1999). It is generally decreased by an increase in water content in the feed and increasing temperature (Smith et al., 1992). Smith et al. (1992) reported that at low specific mechanical energy input levels (SME) the quantity of swollen starch granules increases with increasing SME. Starch granules remain undamaged. Water absorption index (WAI) increases as specific mechanical energy (SME) increases due to an increasing proportion of gelatinized starch granules (Fig. 2.4). The starch solubility also increases with increasing specific mechanical energy (SME) due to macromolecular degradation of starch and especially of amylopectin (e.g. Schuchmann and Danner, 2000). The water solubility and absorption indices reach their maximum at an intermediate level, where damaged and undamaged starch granules are found simultaneously in the melt: appearance and disappearance of gelatinized starch are in balance. As specific mechanical energy (SME) further increases, starch granules are totally damaged. Accordingly, the water absorption index decreases while the water solubility index increases (Fig. 2.4).

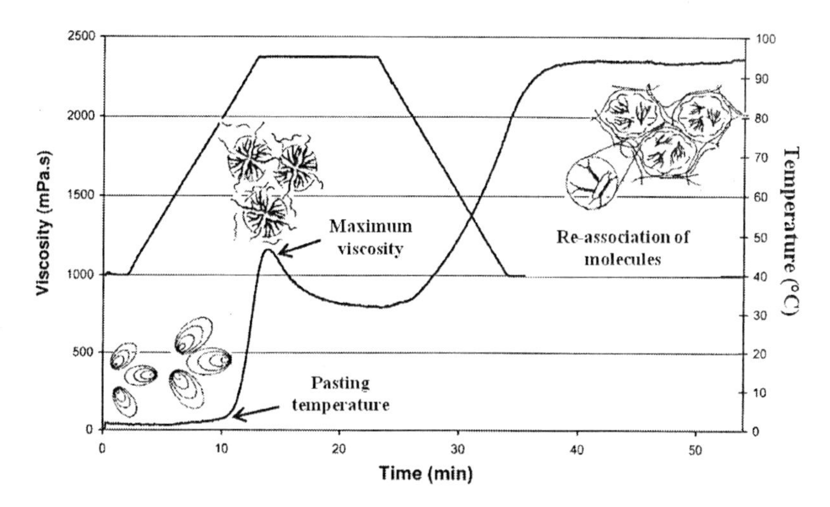

Fig. 2.5: Schematic pasting profile of starch (adapted from Delcour and Hoseney, 2010)

Another test, complementary to the water solubility/absorption indices, used to analyze the transformations of starch during extrusion is based on the degree of gelatinization of starch. Starch gelatinization is an endothermic process which corresponds to the dissociation temperature of the starch molecules from a double helical structure in a starch granule to an amorphous conformation. The gelatinization process is reversible before the temperature reaches the gelatinization temperature. Above this temperature, irreversible processes such as swelling, loss of birefringence and starch solubilization occur (Srichuwong and Jane, 2007). This transition can be determined using a differential scanning calorimeter (DSC) (Srichuwong and Jane, 2007). Gelatinization is also associated with changes in viscosity of the starch paste. Rheometers or viscometers such as the Rapid Visco Analyzer (RVA) are used to measure this change of viscosity. A typical pasting profile of starch is shown in Fig. 2.5. During this measurement taking place under excess of water and increasing temperature, the starch paste viscosity is first increased due to granular swelling and leaching of starch components from the granule. A maximum viscosity is then reached with full swelling of the granule and eventually total disruption of the granule, leading to a decrease in the viscosity of the paste is observed (see Fig. 2.5) (Srichuwong and Jane, 2007). Amylopectin is primarily responsible for granular swelling and viscosity. During granular swelling, the hydrogen bonds between the starch chains are dissociated and replaced with hydrogen bonds with water molecules, resulting in an increase in viscosity (Srichuwong and Jane, 2007). Amylose is the main component of starch leaching out from the granules. The concentration of solubilized amylopectin increases with temperature. During cooling of the hot paste amylose molecules that have leached out from the granules rapidly aggregates. This leads to an increase in the paste viscosity (Srichuwong and Jane, 2007) (see Fig. 2.5). Therefore, changes in the viscosity of the starch paste after extrusion reflects changes in the supramolecular and molecular structure of starch (Srichuwong and Jane, 2007). These changes in the viscosity of the starch paste depend on the extrusion conditions as reported for potato and wheat flour blends by Bhattacharya et al. (1999). The authors showed an increase in paste viscosity when increasing the water content in the extruder from 16 % to 21 %. The same trend was shown when reducing the screw speed from 400 rpm to 200 rpm at high water content. They explained this effect with the reduced degradation of starch granules and molecules. This reduced degradation of starch was due to the reduced shear stress in the extruder when increasing the water content or reducing the screw speed. The gelatinization properties of starch are also affected by the presence of other solutes such as sucrose or sodium chloride (e.g. Chungcharoen and Lund, 1987), hydrocolloids such as gluten or gums (e.g. Christianson et al., 1981; Ghiasi et al., 1982) or fat forming amylose-lipid complexes (e.g. Morrison et al., 1993). The particle size of starchy-flours was shown to significantly influence the starch transformation during extrusion. For instance, Garber et al. (1997) reported reduced starch gelatinization during extrusion when reducing corn flour particle size from 1600 μm to 400 μm.

Most research groups working on cooking extrusion of cereals have reported than starch molecules are degradated during extrusion. Several works have reported that amylopectin is primarily degradated due to its higher hydrodynamic volume (e.g. Davidson et al., 1984; Kingler, et al., 1986; Politz et al., 1994; Willet et al., 1997; Schuchmann and Danner, 2001; Brümmer et al., 2002). Some authors reported a degradation of both amylose and amylopectin during extrusion (Colonna et al., 1984).

II.4.2. Nucleation of bubbles

Following plasticization of starch nucleation of bubbles is expected. Nucleation of bubbles occurs at the very end of the die, leading to the formation of the bubble. Nucleation and bubble growth primarily obey the Laplace equation (2.1). The pressure difference ΔP between pressure of the vapor phase in the interior of the bubble P_V and the pressure of the surrounding liquid phase P_L is given in equilibrium by equation (2.1) (Moraru and Kokini, 2003):

$$\Delta P = P_v - P_L = 2\sigma\left(\frac{1}{R}\right), P_v > P_L \qquad (2.1)$$

σ is the interfacial or surface tension of the liquid-vapor interface and R is the radius of the bubble. During extrusion, nucleation and growth of bubbles occur when the melt pressure at the die reaches the vapor pressure. Nucleation of bubbles consists of the formation of small, thermodynamically unstable, gaseous embryos within the liquid metastable phase.

Once an embryo reaches a critical size, it grows spontaneously into a stable and permanent bubble, called nucleus. As superheating increases, the difference P_V - P_L increases and R becomes larger (see equation 2.1). The bubble approaches molecular dimensions at very high degrees of superheat and becomes visible (Moraru and Kokini, 2003). In the conventional nucleation theories, heterogeneous and homogenous nucleation are distinguished. Both are induced by energy variation in the liquid phase. A maximum energy, depending on the surface tension of the fluid and on the difference between the actual pressure and the vapor pressure is required to obtain a nucleus that grows (Lee, 2000).

Homogeneous nucleation occurs when thermodynamic instability is reached within the liquid phase. In case of homogeneous nucleation, the nucleation rate (number of nuclei per unit volume and per unit time) J_{hom} depends on the melt pressure, the vapor pressures and the temperature according to equation (2.2) (Lee, 2000):

$$J_{\text{hom}} = N_d \left(\frac{2\sigma}{\pi MB} \right)^{\frac{1}{2}} \exp\left(\frac{-16\pi\sigma^3}{3k_B T (P_V - P_L)^2} \right) \tag{2.2}$$

N_d is the atom density in the liquid and M the molecular weight. B equals to $1 - \dfrac{1}{3\left(1 - \dfrac{P_L}{P_V}\right)}$.

k_B is the Boltzmann's constant and T is the temperature.

On the contrary, heterogeneous nucleation occurs at the interface of a liquid and a clean surface (see Fig. 2.7). In this case the nucleation rate J_{het} also takes into account the interfacial properties between the bubble and the surface according to equation (2.3) (Lee, 2000):

$$J_{het} = N_d^{\frac{2}{3}} S_g \left(\frac{2\sigma}{\pi MBF} \right)^{\frac{1}{2}} \exp\left(\frac{-16\pi\sigma^3 F}{3k_B T (P_V - P_L)^2} \right) \tag{2.3}$$

where S_g is a geometric factor that equals to $(1\text{-}p)/2$ and F is another geometric factor defined by equation (2.4):

$$F = \frac{2 - 3p + p^3}{4} \tag{2.4}$$

p = -cos(θ) with θ the contact angle of the liquid with the surface.

For both homogeneous and heterogeneous nucleation the nucleation rate can be modulated by altering the pressure difference between in the inside and the outside of the bubble (P_V- P_L), the interfacial tension σ and the temperature T (Lee, 2000). Often in cooking extrusion, changes in surface tension σ are neglected.

Another type of nucleation involves variation in pressure. It is called *cavitation*. This phenomenon most often occurs when porous surfaces provide cavities for gas molecules. Nucleation can also be induced by the effect of shear stress at the die walls (Lee, 2000). Incorporation of air in the melt through air-filled pores in the granular feed was also reported to be one reason to explain the difference in the degree of nucleation (Hoseney et al., 1992, Cisneros and Kokini, 2002a, b).

II.4.3. Bubble growth

Extrudate expansion is governed by the biaxial extension of individual bubbles. The driving force for bubble growth is the pressure difference between the inside and the exterior of the matrix (see equation 2.1). The rheological properties of the melt have also a leading role. They determine the resistance of the bubble wall to this pressure difference between the inside and the outside of the bubble (Moraru and Kokini, 2003). In order to grow at the die exit, the bubble has to overcome the resistance given by the shear viscosity (Pai et al., 2009). The effect of shear viscosity on the expansion properties have been reported several times (see Chapter II.4.3.3). This can be easily measured using on-line rheology (see e.g. Della Valle et al., 1996). The melt is commonly considered to exhibit a viscoelastic behavior. The importance of the elastic properties of the melt (also called die swell) in the expansion mechanism is very often discussed in literature (e.g. see Bouzaza et al., 1996). Nevertheless measurement of visco-elasticity remains complicated. The use of off-line techniques such as squeeze flow rheology or dynamic mechanical thermal analysis shows limitations for extruded melts. The difficulties in measuring rheological properties with off-line techniques include changes in material physicochemical changes during samples preparation and/or water evaporation during measurement. Therefore, on-line techniques are more appropriate to measure the melt viscosity in the conditions of extrusion. The measurement of shear viscosity with slit viscometers and estimation of extensional viscosity using pressure drop at the die entrance/exit are presented in this paragraph. Literature discussing the relationship between viscosity and expansion properties is also summarized.

II.4.3.1. On-line shear viscosity measurements

The principle of on-line capillary rheometers is based on the measurement of the shear stress in the slit (obtained from the pressure drop ΔP) depending on the shear rate. Single slit rheometers (or viscometer dies) have been extensively used to measure the shear viscosity of extruded starch and flour (see e.g. Senouci and Smith, 1988a). In single slit rheometers, the shear stress in the slit is adapted by changing the extruder feed rate (twin-screw extruders) or the screw speed (single-screw extruders). Changing the feed rate or the screw speed was shown to significantly modify the starch thermomechanical history in the extruder. These changes in process conditions lead to differences in physicochemical properties of starch (e.g. Vergnes and Villemaine, 1987). This may lead to important mistakes when establishing the flow curves whish are showing the shear stress vs. shear rate. In order to study the flow behavior at constant ingredient thermomechanical history, a pre-shearing rheometer ("Rheoplast®, INRA, France") was designed by Vergnes and Villemaire (1987). It enables to reproduce a specific degree of starch transformation in the apparatus and measure the viscosity.

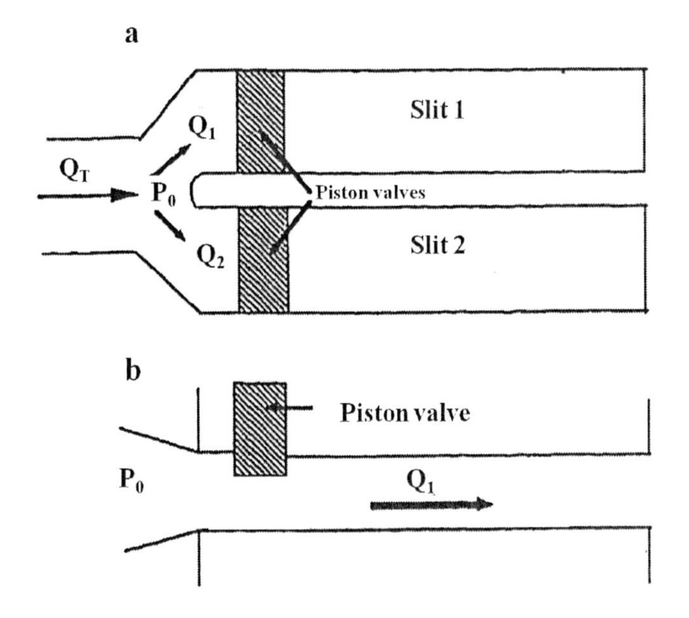

Fig. 2.6: Twin-slit rheometer viewed from above (a) or from the side (b) (Q_T = extruder volumetric rate, $Q_{1/2}$ = volumetric flow rate in left or right slit, P_0 pressure at the rheometer entrance) (adapted from Vergnes et al., 1993)

Twin-slit rheometers ("Rheopac®, INRA/CNRS, France") for food applications were introduced by Vergnes et al. (1993). This rheometer was inspired from the extrusion of polymers and enables to maintain a similar thermomechanical history of the ingredients in the extruder, while changing the shear rate in the slits. The design of this rheometer is displayed in Fig. 2.6. A constant pressure in the extruder is maintained by moving one piston backward and the other one forward. This allows changing the flow rate in the slits (and thus also the shear rate) while maintaining the pressure in the extruder constant.

During viscosity measurement with slit rheometers, slip at the wall may occur. Slip happens when a thin layer of fluid, having a viscosity lower than that of the melt, forms at the wall of the rheometer. To correct this effect, an additional term, representing added flow, is added to the overall flow rate term Q_{ws}. It is calculated according to equation (2.5) (Mooney, 1931; Steffe, 1996).

$$Q_{measured} = Q_m = Q_{ws} + \pi R^2 u_s \qquad (2.5)$$

R is the slit radius in the case of a cylindrical capillary and u_s the effective slit velocity which is assumed to be a function of the wall shear stress σ_w. From the ratio of effective slit velocity to the shear stress at the wall, one can deduct the effective slit coefficient β ($\beta = u_s/\sigma_w$). The slit coefficient is a function of the shear stress. It varies inversely with the tube radius. Therefore a correction is added according to the radius to lead to equation (2.6):

$$\beta_c = \frac{\beta}{R} \qquad (2.6)$$

The corrected volumetric flow rate can then be written according to equation (2.7):

$$Q_{ws} = Q_m - \beta_c \sigma_w \pi R \qquad (2.7)$$

Q_{ws} is then used to calculate the shear rate.

Mathematical fitting of shear viscosity when correlated to process conditions enables to estimate the melt's flow behavior in the extruder and the final expansion properties of extruded foods. Over years several mathematical fittings of shear viscosity as a function of process conditions have been proposed. They are based on data obtained from single or twin-slit rheometer measurement. These mathematical fitting equations integrate the effect of water content and barrel temperature (e.g. Harper et al., 1971 or Cervone and Harper, 1978). Some mathematical fittings of shear viscosity also include the thermomechanical history of the ingredients as it was shown to significantly influence the melt shear viscosity. It is reflected in the different equations by adding the effect of the screw speed (N) (e.g. Senouci and Smith, 1988a), specific mechanical energy (SME) (e.g. Vergnes and Villemaire, 1987) or extruder barrel pressure (P_b) (e.g. Padmanabhan and Bhattacharya, 1991). The most relevant mathematical fittings of shear viscosity depending on process conditions are presented in Tab. 2.1.

Tab. 2.1: Published studies on mathematical fitting of shear viscosity as a function of process conditions (adapted from Martin et al., 2003) (η is the shear viscosity, $\dot{\gamma}$ is the shear rate, n is the power law index, ΔE is the activation energy, T is the melt temperature, MC is the moisture content and P_b represents the barrel pressure in the extruder)

Authors/year	Product	Experimental set-up	Rheological model[a]	Temperature (°C)	Moisture content (%)	Conditions
Harper et al. (1971)	Cereal dough	Single-screw extruder + cylindrical die	$\eta = K_0 \exp(\Delta E/RT)\exp(k\mathrm{MC})\dot{\gamma}^{n-1}$, $\Delta E/R = 2482$ K; $k = -0.079$	67–100	25–30	–
Cervone and Harper (1978)	Corn flour	Single-screw extruder + slit die	$\eta = K_0 \exp(\Delta E/RT)\exp(k\mathrm{MC})\dot{\gamma}^{n-1}$, $\Delta E/R = 4388$ K; $k = -0.101$	90–150	22–30	10–100 rpm, $10^1 < \dot{\gamma} < 10^3\ \mathrm{s}^{-1}$
Fletcher et al. (1985)	Corn grits	Single-screw extruder + slit die rheometer	$\eta = K_0 \exp(\Delta E/RT)\exp(k\mathrm{MC})\dot{\gamma}^{n-1}$, $\Delta E/R = 3969$ K; $k = -0.03$	153–168	15–18	50–200 rpm, $10^1 < \dot{\gamma} < 10^3\ \mathrm{s}^{-1}$
Vergnes and Villemaire (1987)	Corn starch	Pre-shearing rheometer Rhéoplast®	$\eta = K_0 \exp(E/RT - \alpha\mathrm{MC} - \beta W)\dot{\gamma}^{n-1}$, $E/R = 4250$ K; $\alpha = 10.6$; $\beta = 0.088$	110–170	26–49	200–600 rpm, $10^1 < \dot{\gamma} < 10^3\ \mathrm{s}^{-1}$
Senouci and Smith (1988a)	Corn starch	Twin-screw extruder + slit die rheometer	$\eta = K_0 N^{-\alpha} \exp(\Delta E/RT)\exp(k\mathrm{MC})\dot{\gamma}^{n-1}$, $\Delta E/R = 2834$ K; $k = -0.032$; $\alpha = 0.541$	100–140	20–31.5	100–250 rpm, $10^1 < \dot{\gamma} < 10^3\ \mathrm{s}^{-1}$
Padmanabhan and Bhattacharya (1991)	Corn meal	Single-screw extruder + slit die viscometer	$\eta = m\dot{\gamma}^{n-1} \exp(\Delta E/RT + \alpha\mathrm{MC} + gP_b)$, $\Delta E/R = 2726$ K; $\alpha = -1.99$; $g = 3.5 \times 10^{-8}$	150–180	25–45	80–240 rpm, $10^2 < \dot{\gamma} < 10^3\ \mathrm{s}^{-1}$
Willett et al. (1995)	Corn starch	Single-screw extruder + capillary die viscometer	$\eta = K(T_0)\exp[E/RT - \alpha\mathrm{MC}]\dot{\gamma}^{n-1}$, $E/R = 8500$ K; $\alpha = 12.6$	110–130	15–30	1–60 rpm, $10^1 < \dot{\gamma} < 10^3\ \mathrm{s}^{-1}$
Della Valle et al. (1996)	Corn starch (0–70% amylose contents)	Twin-screw extruder + Rheopac	$\eta = K_0 \exp[E/RT - \alpha\mathrm{MC} - \beta\mathrm{SME}]\dot{\gamma}^{n-1}$, $E/R = 6140$ K; $\alpha = 18.6$; $\beta = 2.1 \times 10^{-3}$	100–185	20–36	80–240 rpm, $10^1 < \dot{\gamma} < 10^3\ \mathrm{s}^{-1}$

[a] All variables are expressed in the SI units system (K in Pa s, E/R in Kelvin, β in $(\mathrm{kWh/t})^{-1}$ and g in Pa^{-1}). k and α are dimensionless coefficients.

The mathematical fitting equations are based on the dependency of viscosity (η) over temperature following the Arrhenius relationship (equation 2.8) (Makosko, 1994):

$$\eta = \eta_0 \exp\frac{\Delta E}{RT} \tag{2.8}$$

η_0 is a reference viscosity, ΔE represents the activation energy, R is the gas constant and T is the temperature.

This relationship is only valid for differences between material temperature and glass transition temperature T_g above 100 K ($T > T_g + 100$ K). Therefore the mathematical fittings presented in Tab. 2.1 can only be applied for temperatures above $T_g + 100$ K. These mathematical fitting equations of experimental data are often used to explain the expansion properties of extruded cereals. Nevertheless, expansion is a dynamic process and the expandate leaving the die undergoes evaporating cooling. At lower temperature differences ($T < T_g + 100$ K), the melt properties are better described by the Williams-Landel-Ferry (WLF) equation (Williams et al., 1955) (equation 2.9):

$$\log \frac{\eta}{\eta_g} = \frac{-C_{WLF}\left(T - T_g\right)}{B_{WLF} + \left(T - T_g\right)} \tag{2.9}$$

η and η_g are the viscosity at the temperature T and at the glass transition temperature T_g respectively. B_{WLF} is a dimensionless constant and C_{WLF} is a parameter.

Such an approach was used by Fan et al. (1994) when simulating bubble growth in extruded products. The authors took into account the time-temperature dependency of the melt rheological properties of the extrudate. They reported that a combination of power law model with the Williams-Landel-Ferry (WLF) equation describes well the rheological behavior of an extruded melt.

The only-non Newtonian effect considered in the mathematical fittings of shear viscosity of extruded cereals is the pseudoplastic behavior of the melt (reflected by the power law index n) (see Tab. 2.1). Nevertheless, viscoelasticity may also play a significant role in expansion due to the recovery of the elastic properties at the die exit (Bouzaza, 1996). Some studies incorporate the melt viscoelasticity in rheological models (Brasseur et al., 1998; Dhanasekharan & Kokini, 2000). These studies concluded that the viscoelastic effect responsible for the leakage flow in the extruder is minimal. This supports that using simple viscosity models to characterize melt rheology in the extruder is acceptable (Moraru & Kokini, 2003).

II.4.3.2. Measurement of die entrance pressure drop

Starch melts is usually considered to exhibit a viscoelastic behavior. The measurement of the elastic component of molten plasticized starch associated with the first normal stress difference (N_1) is challenging. Indeed it is difficult to reproduce the extrusion conditions in a conventional rheometer (Martin et al., 2003). The exit and entrance pressure drop measured in on-line capillary/slit rheometers have been used by several authors to estimate the elastic properties of starchy melts (Senouci and Smith, 1988b; Padmanbahn and Bhattacharya, 1991; Chang, 1992; Ofoli and Steffe, 1993; Martin et al. 2003). Restriction of the melt at the entrance of the rheometer/die accelerates the flow. The flow is therefore not steady anymore. This acceleration of the flow enables measuring the elastic properties of the melt. Indeed elastic work must be done to deform the macromolecules to their strained state corresponding to the high stress inside the restricted flow part. This work is provided through an extra pressure drop at the entrance (Padmanbahn and Bhattacharya, 1991).

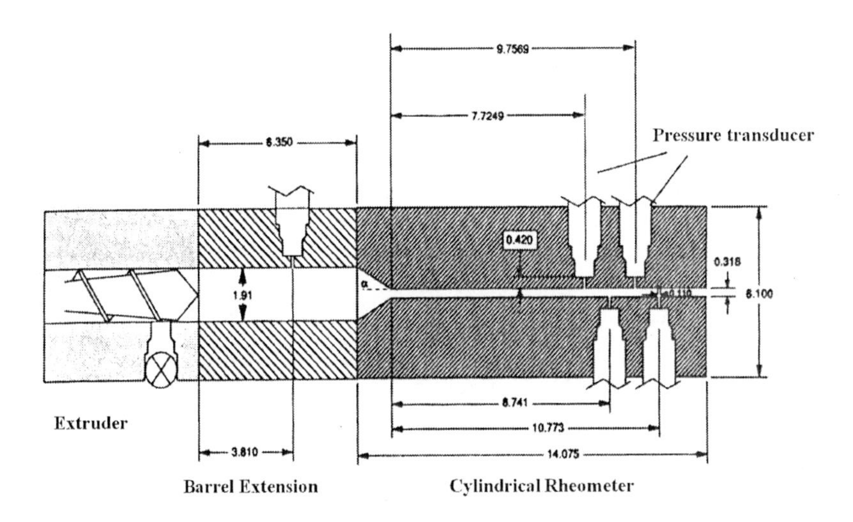

Fig. 2.7: Schematic illustration of an extruder/capillary die arrangement to measure the pressure drop at the entrance (adapted from from Bhattacharya et al., 1994)

A schematic representation of the extruder/capillary rheometer set-up used to measure the entrance and exit pressure drops is shown in Fig. 2.7. In on-line capillary rheometry, a large variation of pressure drop is associated with the change in flow velocity at the entrance and exit regions (Fig. 2.8). This drop in melt pressure at the entrance and exit of the rheometer can be explained by shear and extensional forces. The total pressure drop in the slit ΔP is composed of the pressure drop over the length of the capillary (flow fully developed) ΔP_{Cap} and pressure drop at both ends ΔP_{End} ($\Delta P_{End} = \Delta P_{Entrance} + \Delta P_{Exit}$) (Fig. 2.8). ΔP is then calculated according to equation (2.10):

$$\Delta P = \Delta P_{End} + \Delta P_{Cap} \qquad (2.10)$$

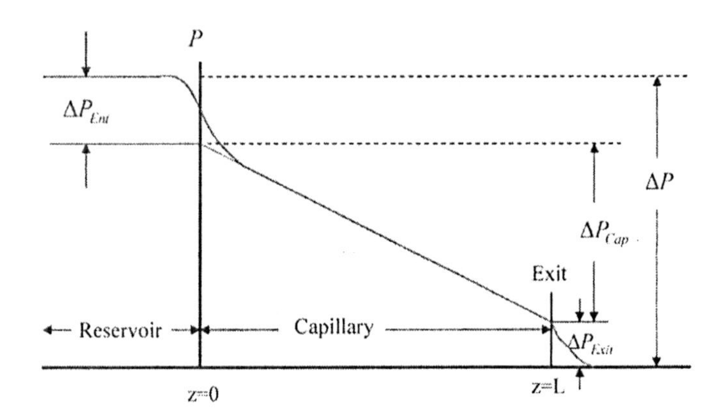

Fig. 2.8: Schematic diagram of the pressure losses in a flow through a capillary rheometer (ΔP, ΔP_{Ent}, ΔP_{Cap}, ΔP_{Exit} are the total pressure drop, pressure drops at the entrance, along the capillary and at the exit, respectively) (adapted from Mitsoulis and Hatzikiriakos, 2003)

ΔP_{End} is the excess pressure drop due to the entrance and exit flows. Bagley (1957) suggested an empirical method to determine ΔP_{End}, named Bagley correction or end correction n_B:

$$n_B = \frac{\Delta P_{End}}{2\sigma_w} \qquad (2.11)$$

σ_w is the wall shear stress in the die for fully developed flow. In the capillary die, the wall shear stress can be calculated according to equation (2.12):

$$\sigma_w = \frac{R}{2}\frac{dP}{dz} = \frac{1}{2}\frac{\Delta P}{n_B + L/R} \qquad (2.12)$$

R and L are the radius and length of the capillary, respectively. By plotting ΔP against L/R (called Bagley plot) and extrapolating to $\Delta P = 0$, the Bagley correction n_B can be determined. The end pressure ΔP_{End} can then be obtained from equation (2.10). From the end pressure, the apparent extension rate $\dot{\varepsilon}$ and apparent extensional viscosity η_{pa} can be measured (Mitsoulis and Hatzikiriakos, 2003). Several analyses are available to determine the extensional flow properties. The method of Cogswell (1972) is a simple method based on the balance of the pressure drops between the end pressure and capillary pressure drop. The advanced method of Binging (1988) is an approximation method using variational energy principles. Both methods are based on assumptions. They were reported to reveal considerable differences in their predictions (Mitsoulus and Hatzikiriakos, 2003).

The data points in the Bagley plot do not always lie on straight lines and therefore extrapolation is not always possible and straightforward (Mitsoulis and Hatzikiriakos, 2003). Therefore, some authors used the orifice die method. This method consists in extrapolating the value of the entrance pressure drop ΔP_{Ent} to $L/D \approx 0$ (corresponding to a virtual zero die length). This extrapolation enables obtaining the pressure drop at the die entrance associated to extensional properties of the flow. This method was shown by Hatzikiriakos and Mitsoulis (1996) to lead to comparable results to those obtained with Cogswell (1972).

Senouci and Smith (1988b) used the Cogswell correction (Cogswell, 1972) to evaluate the extensional viscosity η_{pa} of extruded maize grits and potato powder or wheat starch (in case of a planar contraction) from the entrance pressure drop ΔP_{Ent} according to equation (2.13):

$$\eta_{pa} = \frac{3(n+1)^2 \Delta P_{Ent}^2}{4\dot{\gamma}_a^2 \eta_a} \qquad (2.13)$$

n is the power law index of shear viscosity, $\dot{\gamma}_a$ is the apparent shear rate and η_a is the apparent shear viscosity at the wall. The authors reported a decrease in extensional viscosity with shear rate and significant differences in extensional viscosity depending on the cereal source (Senouci and Smith, 1988b). Martin et al. (2003) used a similar approach for wheat flour. The authors reported a decrease in extensional viscosity with shear rate. Chang (1992) used the exit pressure (P_{Exit}) method with a slit rheometer. The author estimated the normal stress difference N_1 during starch extrusion using several assumptions and according to the equation given by Han (1974) (equation 2.14):

$$N_1 = P_{Exit} + \left(\frac{\partial P_{Exit}}{\partial \sigma_w} \right) \qquad (2.14)$$

σ_w is the shear stress at the wall of the rheometer and P_{Exit} is the exit pressure. The author observed that the first normal stress difference increases with shear rate and decreases with material temperature and water content. Martin et al. (2003) reported a similar effect of shear rate on the normal stress difference of extruded wheat flour. Nevertheless, the validity of the exit pressure method has been questioned because of flow disturbances, velocity rearrangements or extrudate expansion may affect the shear near the die exit, thus invalidating the developed flow assumption (Martin et al., 2003). Pai et al. (2009) used squeezed flow rheology to measure the extensional viscosity of extruded corn flour. They reported an increase in extensional viscosity when increasing the corn bran content.

II.4.3.3. Relationship between melt viscosity and expansion

The growth of the cells and bulk expansion properties are closely linked. They are generally characterized by measuring the volumetric expansion index (VEI), the sectional expansion index (SEI) and the longitudinal expansion index (LEI). The SEI reflects the expansion of the melt in the sectional direction of the flow. LEI characterizes the expansion of the melt in the direction of the extrusion flow. The effect of the process conditions on the expansion indices has been reported several times (see Moraru and Kokini, 2003). The effect of the process conditions on the expansion properties can be described by a quadratic relationship with a maximum which corresponds to an optimal melt viscosity (see Chinnaswamy, 1993). The rheological properties determine the resistance of the bubble wall to the pressure difference between the inside and the outside of the bubble (Moraru and Kokini, 2003). Thus they are decisive for the expansion. Several authors have tried to link the shear viscosity of the melt, assessed with single or twin-slit rheometers, with the expansion properties (e.g. Della Valle et al., 1997). Some workers applied simple mathematical fitting of the experimental values. Using a single slit rheometer, Alvarez-Martinez et al. (1988) established a relationship between the shear strain at the die, the process conditions and the expansion properties. They reported that the sectional expansion was decreased by an increase in moisture and an increase in shear strain at the die. The longitudinal expansion was decreased both by temperature of the melt and shear strain at the die. The authors fitted the experimental points using an Arrhenius-type of relationship as described in equation (2.15):

$$Y = \exp\left(a + bMC\right) \exp\left[\frac{\Delta E}{RT}\right] \left(\bar{\gamma}_d\right)^f \left(\bar{\gamma}_{ms}\right)^g \qquad (2.15)$$

Y is the expansion index (volumetric, sectional or longitudinal), MC is the water content, T is the temperature, ΔE is the activation energy, R is the gas constant, $a/b/d/f/g$ are constants and $\bar{\gamma}_d$ and $\bar{\gamma}_{ms}$ are respectively the total shear strain at the single-hole die and in the measurement section of the screw.

Bouzaza et al. (1996) reported an increase in sectional and longitudinal expansion indices with the ratio of pressures difference between the vapor pressure in the bubble and the atmospheric pressure ΔP to the apparent shear viscosity of the membrane surrounding this bubble $(\mu_a)_d$. The measurements were performed at constant die diameter. This relationship was used for the fitting of the experimental data according to equation (2.16). This relationship showed a logarithmic relationship between expansion (index Y) and $\Delta P/(\mu_a)_d$.

$$\mathrm{Y} = K_e D_d^{\alpha} L_d^{\beta} \ln\left(\frac{\Delta P}{\left(\mu_a\right)_d} - C\right) \qquad (2.16)$$

L_d and D_d are the die diameter and length, respectively and α, β, K_e and C are constants. When changing the die diameter, the authors observed an inverse relationship between the sectional and longitudinal expansion index (Bouzaza et al., 1996). They attributed this effect to the elastic properties of the melt. When the elastic properties of the melt are favored, the longitudinal expansion is no longer favored, explaining their inverse relationship.

Della Valle et al. (1997) described the relationship between the apparent melt shear viscosity, obtained from a twin-slit rheometer, and the expansion properties at the rheometer slits exit. The authors reported an increase in volumetric expansion when the apparent shear viscosity was decreased. This effect was obtained when increasing shear rate at constant temperature, water content or specific mechanical energy. The importance of extensional viscosity (or elongational viscosity) on expansion properties of extruded corn flour with added corn bran was recently highlighted by Pai et al. (2009). The authors reported that once the bubble has reached a significant size, the membrane surrounding the bubble undergoes biaxial extension. As the bubble is growing, the membrane surrounding the bubbles is thinning. Once a certain normal stress in the melt membrane has been exceeded due to further expansion, the membrane around the bubbles ruptures causing bubbles to collapse. For bubbles to remain for the same expanded size, the melt should have an extensional viscosity high enough to withstand the extensional stress caused by bubble expansion.

II.4.4. Bubble growth cessation and collapse

Expansion is a dynamic process. During bubble growth, the temperature drops due to the evaporative cooling. The extrudate reaches a glassy state after crossing the rubbery region and expansion stops (Moraru and Kokini, 2003). The temperature at which the bubble growth ceases corresponds to a viscosity for which the mobility of the molecules is not any longer enough to further expand. This critical difference between the melt temperature and the glass transition temperature T_g is debated in the literature, especially as the glass transition is not a sharp transition. For instance a temperature value of T_g + 30 K, often used a reference, was given by Fan et al. (1994). In order to estimate the physical changes of the material during extrusion and later during expansion according to the temperature and water content, a state diagram showing the phase transitions (e.g. melting and glass transition temperature depending on the water content) can be built. Such state diagrams, based on DSC measurements, were built for wheat flour components by Kokini et al. (1994), Kaletunç and Breslauer (1996) and later by Cuq et al. (2003). By plotting the hydrothermal path of the samples depending on their conditions of temperature and water content during extrusion, the change in phase transitions can be compared depending on the composition of the extruded material.

During bubble growth, the bubble wall may not withstand the pressure inside the bubble, leading to collapse. The collapse may have a significant effect on the extrudate final dimensions (Moraru and Kokini, 2003) (also see Fig. 2.3). Modeling of bubble growth and collapse demonstrated that collapse is higher at low extrusion temperature, while it is insignificantly higher at high temperatures (Fan et al., 1994). Low shear viscosity favors high collapse, while elastic properties may also play a role according to the extrusion conditions (Moraru and Kokini, 2003).

II.5. Extrusion of composite materials: Inclusion of dietary fibers in extruded cereals

The effect of fibers on the expansion properties and final texture and taste of extruded products depend on the type and concentration of fibers. Adding an increasing concentration of insoluble fibers such as sugar beet fibers (Hsieh et al., 1991; Lue et al., 1991), wheat fibers (Hsieh et al., 1989; Yanniotis et al., 2007), oat fibers (Hsieh et al., 1989) or cereal bran such as wheat bran (Breen et al., 1977; Moore et al., 1990; Wang et al., 1993; Onwulata et al., 2001; Brennan et al., 2008) or corn bran (Blake, 2006) to extruded cereal flours was shown to result in a significant reduction of the samples sectional expansion and in a significant increase in their bulk density. This also resulted in an increase in their modulus of deformability and breaking stress (Hsieh et al., 1989; Moore et al., 1990; Wang et al., 1993). The reduction in expansion produces denser, though and non-crispy products (Lue et al., 1991). Microscopy observations, followed in some studies by image analysis, showed that incorporation of wheat bran yields structure with an increased cell density and reduced cell sizes (Moore et al., 1990; Lue et al., 1990; Yanniotis et al., 2007).

Incorporation of soluble fibers such as psyllium and inulin in extruded starchy products were reported to produce a better expansion than insoluble fibers (see e.g. van Lengerich & Larson, 2000). For instance, Brennan et al. (2008) showed no significant changes in the sectional expansion of white flour with added inulin to 15 %. Nevertheless the density of the samples with added inulin was significantly increased. The cost of soluble fibers, their tendency to reduce the crispiness of products and dissolve in water producing a gummy or slimy product limits the quantity that can be incorporated in extruded breakfast cereals (Blake, 2006). The effect of hydrocolloids such as carboxymethylcellulose, acacia gum or sodium alginate on rice grits was also investigated. They were reported to lower the expansion (Kaur et al., 1999).

The effect of wheat bran or wheat fibers on the melt rheological properties was rarely studied up to date. Moore et al. (1990) and Wang et al. (1993), using an on-line capillary rheometer, reported no significant change in the apparent shear viscosity of extruded wheat flour supplemented with up to 16 % of wheat bran. Guy (1985), Moore et al. (1990) and Yanniotis et al. (2007) attributed the reduction in expansion of wheat bran-containing products to the interference of bran with the bubble expansion. The low adhesion properties between the bran particles and the bubble membrane reduced the extensibility of bubble walls and caused premature rupture of steam bubbles. The rupture was favored at a critical bubble wall thickness equal to the particle size of bran. Pai et al., (2009) associated the decrease in expansion when increasing the corn bran content in corn starch with both an increase in the extensional viscosity and shear viscosity.

The effect of fiber or cereal bran particle size was shown to affect expansion properties. Lue et al. (1991) reported an increase in both sectional and longitudinal expansion of corn meal when decreasing the particle size of sugar beet fibers from 200 µm to 10 µm. They attributed this effect to a reduction in the degree of bursting of the bubbles at lower particle size and to the higher water-binding capacity of sugar beet fibers with low particle size. Guy (1985) noted an increase in volumetric expansion and sectional expansion when decreasing the wheat bran particle size from 1500 µm to 150 µm.

As earlier reported only few scientific studies investigated the effect of insoluble fibers on the viscoelastic properties of extruded cereals. More literature is available in the field of synthetic polymers reinforced with synthetic fibers. The effect of synthetic fibers on the viscoelastic properties of synthetic polymers is depending on their density, dimensions (size and aspect ratio), orientation and surface interactions with the continuous polymer phase (Xanthos, 2005). They were reported to increase the shear viscosity and pseudoplastic flow behavior of synthetic polymers and decrease their elastic properties (Xanthos, 2005).

II.6. Mechanical properties of solids foams

II.6.1. Relationship between instrumental and sensory measurements

The texture of puffed foods is one of the main parameters driving consumer preference. It is a product fingerprint that differentiates one competitor to another. Among the attributes of puffed foods, that of crispiness has been a major focus for many researchers. A scientific paper reviewed the mechanisms underlying crispiness perception and the tools used to assess it (Roudaut et al., 2002). Among the different techniques to determine the textural properties of puffed foods, instrumental measurements are the most used especially by the Food Industry. They offer fast, easy and cost-efficient solutions to access the mechanical properties. For some of them close correlation with in-vivo sensory attributes such as crispiness were reported by e.g. Vallés Pàmies et al. (2000).

Crispiness is defined as a combination of sounds and mechanical properties (Roudaut et al., 2002). Large deformation and fracture tests appear to be the most suitable instrumental test to measure the component of crispiness associated to its mechanical properties. Nevertheless, small deformation data, such as those acquired in dynamic rheology may provide information on the molecular level concerning the crispiness attributes (Roudaut et al., 2002). Among the most commonly used instrumental tests, flexure, shear and compression tests are differentiated. The compression test is probably the most commonly used because of its similarities with the mastication process. In this test, the sample is compressed either between two parallel plates or by a plunger compressing the sample. Samples can be tested individually or as bulk when contained in a cell. From this test, several types of stress-strain compression curves can be differentiated according to the type of material and structure tested. Solid foams show a linear elasticity followed by a long collapse plateau, truncated by a regime of densification in which the stress rises steeply (Gibson and Ashby, 1997). The different regimes of linear elasticity, collapse and densification according to the type of foam are shown in see Fig. 2.9.

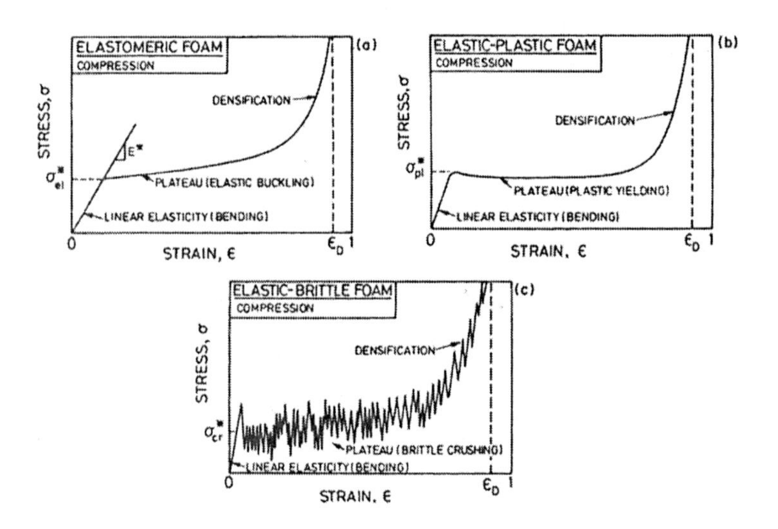

Fig. 2.9: Schematic stress (σ)-strain (ε) compression curves of different foams showing the regimes of linear elasticity, collapse and densification: (a) elastomeric foams, (b) elastic-plastic foam and (c) elastic-brittle foam (adapted from Gibson and Ashby, 1997)

The data analysis of the stress-strain relationship was primarily focusing on the linear region of the force-deformation plot, reflecting the mechanical properties. Another approach is to collect information from the jagged part of the force-deformation curves obtained when using a plunger (see Fig. 2.11c). Using fractal analysis of this jagged part Barrett et al. (1994) described the relationship with the foam cell area A and the bulk density ρ according to equation (2.17):

$$\text{Fractal dimension} = 1.37 - 0.0112A + 1.93\rho \quad (2.17)$$

Puncture tests have been extensively used as well as they mimic the incisors impact at biting (e.g. Ding et al., 2005; Agbisit et al., 2007). This test has been employed for the characterization of foamed products. The probe is expected to fracture, one after the other, the different cell walls constituting the product (Roudaut et al., 2002). Nevertheless, this implies that the diameter of the plunger is lower than the diameter of the cells. Flexure tests have also been extensively used to measure the mechanical properties of the bulk sample structure (e.g. Warbuton et al., 1990; Babin et al., 2007; Robin et al, 2010a). From the linear region in the force-deformation plot, the stress at rupture and elastic modulus of the bulk structure can be calculated and related to the sample relative density according to the Gibson-Ashby models (Gibson and Ashby, 1997), explained in Chapter II.6.2.

The composition of the cell wall material, surrounding the air bubbles, and its transformation occurring during processing may affect the mechanical properties of solid foams. For instance, several authors reported that crispiness is affected by hydration. They attributed this observation to the reduction in glass transition temperature below ambient temperature and thus to a decrease in the rigidity of the hydrated product (e.g. Ablett et al., 1996). Other small molecules such as sucrose may also affect the cell wall mechanical properties.

II.6.2. The Gibson-Ashby model for solid foams

At same shape and dimensions, the mechanical properties of a solid foam depend, above all, on its relative density: the ratio between the bulk density of the foam ρ^* and the density of the cell walls ρ_s. The relative density is obtained from the cell and cell wall dimensions but differs if the cells are open or closed (Fig. 2.10) (Gibson and Ashby, 1997).

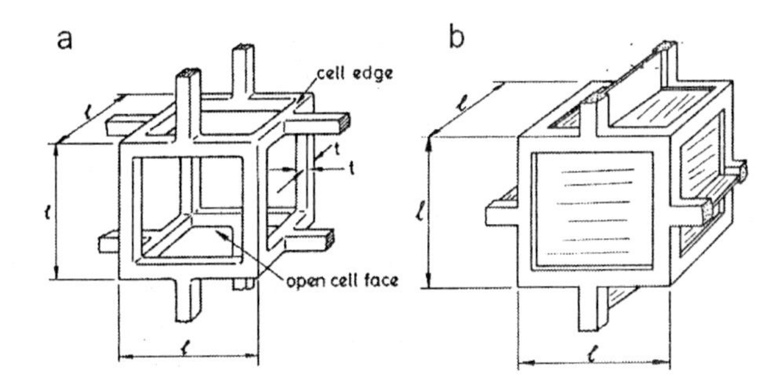

Fig. 2.10: Cubic cell model for an open-cell foam (b) and a closed-cell foam (b) (*l* is the edge length and *t* is the cell wall thickness) (adapted from Gibson and Ashby, 1997)

Based on packing of cubic cells, for open-cell foams with edges of length l and thickness t, the relative density is given by equation (2.18) (see also Fig. 2.10a) (Gibson and Ashby, 1997):

$$\frac{\rho^*}{\rho_s} = C\left(\frac{t}{l}\right)^2 \qquad (2.18)$$

ρ^* is the bulk density of the foam, ρ_s is the density of the cell walls and C is a constant.

For closed-cell foams with faces of side l and uniform thickness t the relative density is calculated according to equation (2.19) (see also Fig. 2.10b) (Gibson and Ashby, 1997):

$$\frac{\rho^*}{\rho_s} = C_1\frac{t}{l} \qquad (2.19)$$

In the domain of linear elasticity, the Young modulus of solid foams E^* is obtained from the slope between the stress σ and strain ε giving equation (2.20) (Gibson and Ashby, 1997):

$$E^* = \frac{\sigma}{\varepsilon} \qquad (2.20)$$

Its relationship with the foam relative density is given in the case of open-cell structures by equation (2.21) (Gibson and Ashby, 1997):

$$\frac{E^*}{E_S} = C_2\left(\frac{\rho^*}{\rho_s}\right)^2 \qquad (2.21)$$

For closed cells, the fraction ϕ of the solid contained in the cell edges must be taken into account according to equation (2.22) (Gibson and Ashby, 1997):

$$\frac{E^*}{E_S} = C_2\phi^2\left(\frac{\rho^*}{\rho_s}\right)^2 + C_2^{'}(1-\phi)\frac{\rho^*}{\rho_s} \qquad (2.22)$$

In the case of open-cell structures, the brittle crushing stress at rupture σ_{cr}^* of the solid foam can be related to its relative density according to equation (2.23) (Gibson and Ashby, 1997):

$$\frac{\sigma_{cr}^*}{\sigma_S} = C_3\left(\frac{\rho^*}{\rho_s}\right)^{3/2} \qquad (2.23)$$

And for closed cells according to equation (2.24) (Gibson and Ashby, 1997):

$$\frac{\sigma_{cr}^*}{\sigma_S} = C_3\left(\phi\frac{\rho^*}{\rho_s}\right)^{3/2} + C_3'(1-\phi)\frac{\rho^*}{\rho_s} \qquad (2.24)$$

Experimentally, it has been found that the relative elastic modulus and stress at rupture are increased with the foam relative density according to the Gibson-Ashby model given by equations (2.25) and (2.26), respectively (Gibson and Ashby, 1997):

$$\frac{\sigma_{cr}^*}{\sigma_s} \propto \left(\frac{\rho^*}{\rho_s}\right)^n \qquad (2.25)$$

$$\frac{E^*}{E_s} \propto \left(\frac{\rho^*}{\rho_s}\right)^m \qquad (2.26)$$

For open cells, $n = 1.5$ and $m = 2$, and for closed cells $n = 2$ and $m = 3$.

The relationships given in equations (2.25) and (2.26) between the relative density and the relative mechanical properties are very often used to characterize starchy foam and determine their cell connectivity (see e.g. Warburton et al., 1992; Lourdin et al., 1995; Trater et al., 2005; Babin et al., 2007; Robin et al., 2010a). Some authors also used equations 2.21 and 2.23 to estimate the fraction ϕ of the solid contained in the cell edges (also called drainage factor) from experimental data (e.g. Agbisit et al., 2006). The cell size and cell wall thickness distribution was also reported to influence the mechanical properties (e.g. Babin et al., 2007).

II.7. Conclusions and scientific gaps

Incorporation of wheat bran reduces the expansion properties of extruded starchy products. Wheat bran is therefore affecting the parameters driving the expansion of solid starchy foams and their resulting mechanical properties. Thus it is of high scientific interest to indentify and quantify the parameters that are modified by the addition of bran. This will enable to provide solutions for an optimal design of the texture of extruded products containing bran. A large amount of literature on the extrusion of pure starch or flour is available. However, literature dealing with complex products containing minor ingredients such as fibers and especially wheat fibers is limited.

The expansion properties of starchy foams are influenced by physicochemical properties of starch and its transformations during extrusion. The effect of wheat bran on the degree of starch transformation has never been addressed. This change in degree of starch transformations may be caused by changes in the starch melting and glass transition temperatures due to competition for water, modification of the rheological properties of the melt and/or protection of starch granules against shear.

Inclusion of bran in extruded starchy melts may modify the viscoelastic properties of the melt at the die exit. This may be due to differences in mechanical properties between the bran particle and the continuous starch matrix. Competition for water or interactions at the interface between starch and bran may also modify the viscoelastic properties of the melt. Published studies reported no change in the shear viscosity of starch when incorporating bran to 16 % (see Chapter II.5). This is surprising considering the increase in shear viscosity of synthetic polymers when supplemented with synthetic fibers (see Chapter II.5). This may indicate a threshold bran concentration from which the shear viscosity starts to change. A single slit/capillary rheometer was used in these studies. Nevertheless, it does not enable to characterize the melt flow behavior. On the opposite, on-line twin slit rheology allows characterizing the behavior of the flow while maintaining a similar thermomechanical input to the ingredients. To our knowledge, it has never been applied to matrices containing wheat bran. The changes in the elastic properties of starchy products induced by the addition of bran have been very often mentioned but only a few times measured (see Chapter II.4). Additionally, in the few relevant studies, the elastic properties of the melt were assessed using off-line techniques. On-line techniques offer the advantage of measuring the rheological properties at the conditions of extrusion. It avoids further modifications of the samples during preparation and analysis. The effect of wheat bran on the expansion properties of extruded starchy foams has been attributed so far to a reduction in the extensibility of bubbles and early burst of the bubbles (see Chapter II.5). Other hypothesis may be applied to explain these reduced expansion properties. They may involve increased nucleation at the die influencing the rheological properties at the die exit, increase in the shear viscosity at constant shear rate creating more resistance against growth and/or increased shrinkage of the structure.

Several studies have reported the effect of bran on the mechanical properties of starchy foams (see Chapter II.5). The authors correlated the mechanical properties with the bulk dimensions of the samples (e.g. diameter and/or porosity). Nevertheless, to our knowledge no work has reported the effect of the cellular structure, generated by the addition of bran, on the mechanical properties of starchy foams with similar porosities. Thus this is of key importance in understanding the effects of bran on expansion and resulting product properties. Indeed for similar porosities, some cellular structure characteristics may lead to preferred textures by consumers compared to others. Understanding texture on a microscopic level requires an accurate characterization of the cellular structure. 3D X-ray imaging techniques have been developed and applied to food structures in the recent years (Babin et al., 2007; Robin et al., 2010a). They enable a more accurate and reliable quantification of the cellular structures compared to older techniques (e.g. two dimensional imaging) applied in past studies of wheat-based cellular structures.

The scientific gaps highlighted in this paragraph will be addressed in this work.

CHAPTER III. MATERIAL AND METHODS

III.1. Material

Wheat flour type 550 and wheat bran were supplied by Provimi Kliba S.A. (Cossonay, Switzerland). Wheat bran (hereafter also mentioned as "bran") was added to the refined wheat flour (hereafter coded RF) at different concentration. For statistical planning reasons two levels of dietary fiber content were chosen: a minimum level of 12.6 % (hereafter coded LB for low bran) and a maximum level of 24.4 % (hereafter coded HB for high bran). The composition of the recipes is shown in Tab. 3.1. The effect of particle size of bran on microstructure and product properties was investigated by using two qualities of wheat bran: a "fine" bran (see Tab. 3.2) with an average volume weighted particle diameter $D_{[4,3]}$ of 224 µm ± 6 µm and a "coarse" bran (see Tab. 3.2) with an average volume weighted particle diameter $D_{[4,3]}$ of 317 µm ± 1 µm. The particle size distribution of the refined wheat flour, fine and coarse bran flours is shown in the Appendices in Fig. 10.1.

Table 3.1: Samples composition

	Starch[1] [%, w/w, d.m.]	Proteins[2] [%, w/w, d.m.]	Fat[3] [%, w/w, d.m.]	Fibers[4] [%, w/w, d.m.]	Ash[5] [%, w/w, d.m.]
Refined flour (RF)	78.5 ± 0.1	13.0 ± 0.2	1.1 ± 0.0	2.8 ± 0.2	0.8 ± 0.0
Low bran (LB)	69.7 ± 0.6	13.9 ± 0.5	1.6 ± 0.0	12.6 ± 0.2	1.8 ± 0.0
High bran (HB)	55.5 ± 1.5	15.0 ± 1.0	2.3 ± 0.1	24.4 ± 0.2	3.2 ± 0.1
Wheat bran (WB)	17.8 ± 0.2	15.6 ± 0.0	3.9 ± 0.0	51.4 ± 0.2	6.6 ± 0.0

[1] Obtained by difference from the other flour components, [2] Measured using AOAC 997.06, [3] Measured using AOAC 983.23, [4] Measured using AOAC 985.29, [5] Measured using AOAC 923.03 (AOAC: Association Of American Chemists)

III.2. Extrusion experiments

A co-rotating double screw extruder (Evolum 25, Clextral, Firminy, France) with a barrel length L_e of 400 mm and a screw diameter D_e of 25 mm was used to process the samples. This results in a ratio L_e/D_e of 16. The screw configuration was composed of conveying and mixing elements with reverse elements (pushing flow in the opposite direction to the travel of the punch) in the last section of the barrel. The product pressure P_d and temperature T_d were measured in the front plate, after the fourth barrel zone and before the die entrance, using a pressure sensor (Kistler 4090B, Kistler Instrument A.G., Winterthur, Switzerland) and a thermocouple (Type J, ROTH+CO. A.G., Oberuzwil, Switzerland). The extruder was operated at a constant feed rate of 10 kg h^{-1}. The three first heating zones of the barrel were kept at a temperature of 40 °C, 80 °C and 120 °C, respectively. The screw speed, temperature of the fourth barrel zone and water content in the feed were varied according to the experimental plans described in Tab. 3.2. Wheat flour was mixed with wheat bran prior to extrusion in a powder mixer (Prodima MJ50, Mecatex/Prodima, St Sulpice, Switzerland) and fed into the extruder using a co-rotating twin-screw feeder (K-Tron Co, Niederlenz, Switzerland). Water (temperature of 20 °C ± 2 °C) was injected into the extruder at a controlled flow rate with a syringe pump (Teledyne ISCO 500D, Teledyne Isco Inc. Lincoln, Nebraska, USA).

The Specific Mechanical Energy (SME, kJ kg^{-1}) was calculated according to equation (3.1):

$$\mathrm{SME} = \frac{\frac{n_{act}}{n_{\max}} \times M - \frac{n_{act}}{n_{\max}} \times M_{unload}}{m_{total}} \times P_{\max} \qquad (3.1)$$

M and M_{unload} are the motor torque (in Nm) under load and without load, n_{act} and n_{max} are the actual and maximum screw speed (rpm), m_{total} is the mass flow rate (kg h^{-1}) and P_{max} is the maximum engine power, which is 27 kW. A circular die of 10 mm length and 3.2 mm diameter was used. The extruded samples were collected and dried in an oven at 60 °C for 16 hours.

For the measurement of the pressure drop at the entrance of a die, another extruder was used in the facilities of the Karlsruhe Institute of Technology (Germany). A co-rotating double screw extruder (ZSK 26 Mc, Coperion/Werner & Pleiderer, Stuttgart, Germany) with a barrel length L_e of 700 mm and a screw diameter D_e of 25.5 mm was used (L_e/D_e ratio of 27.5) was used for this purpose. The screw configuration was composed of conveying and mixing elements with reverse elements in the last sections of the barrel. The product pressure P_d and temperature T_d were measured and before the die entrance, using a pressure sensor (Kistler 4090B, Kistler Instrument A.G., Winterthur, Switzerland) and a thermocouple (PT 100, ROTH+CO. A.G., Oberuzwil, Switzerland). The extruder was operated at a constant feed rate of 10 kg h^{-1}. The temperature profile in the six temperature-controlled zones of the barrel was T_1 = 40 °C, T_2 = 40 °C, T_3 = 70 °C, T_4 = 120 °C, T_5 = 120 °C and T_6 = 120 °C, respectively. The screw speed was 600 rpm and the water content in the feed was 20 %.

III.3. Experimental planning

An experimental design was used in this study to investigate the main parameters determining the changes of the product properties. The use of experimental planning enables to obtain the main effect of each process parameters. This is obtained by running a limited number of trials. The experimental design also provided information on the interacting effects between the bran concentration and the process conditions. The study was focusing on the trends (e.g. decrease or increase) between the process parameters and the measured product properties. For each process parameter, a minimum and a maximum value were chosen based on preliminary experiments and past studies (Robin et al., 2010a). The lower and upper limits of process parameters were chosen to run the extruder in stable conditions and with no blockage. A full factorial experimental plan of 8 experiments (numbered hereafter from 1 to 8) was designed to measure the effect of water content in the feed (18 % and 22 %), barrel temperature (120 °C and 180 °C) and screw speed (400 rpm and 800 rpm) on properties of the extruded refined wheat flour extrudates (RF in Tab. 3.2). A second factorial fractional experimental plan consisting of 16 experiments was designed to investigate, additionally to the barrel temperature, feed water content and screw speed, the effect of bran particle size (fine and coarse) and fiber concentration (12.6 % and 24.4 %) (LB and HB, Tab. 3.2). The effect of the process conditions (screw speed, barrel temperature and water content in the feed), bran particle size and fiber concentration was studied on 14 variables:

- Specific mechanical energy (SME),
- Water absorption index (WAI),
- Water solubility index (WSI),
- Volumetric expansion index (VEI),
- Maximum sectional expansion index (SEI_{max}),
- Sectional expansion index (SEI),
- Longitudinal expansion index (LEI),
- Shrinkage ratio (SR),
- Mean cell size (MCS),
- Mean cell wall thickness (MCWT),
- Cell density (N_c),
- Shear viscosity at 30 s^{-1} ($\eta(30\ s^{-1})$),
- Consistency factor (K)
- Flow power law index (n).

A total of 24 experiments were conducted.

Tab. 3.2: Experimental plans with varying process parameters, dietary fiber content and bran quality levels (RF: refined flour, LB: low bran concentration, HB: high bran concentration)

Samples	Barrel Temp. [°C]	Water content in the feed [%, w/w, w.b.]	Screw speed [rpm]	Bran particle size [-]	Fibers content [%, w/w, d.m.]
RF1	120	18	400	NA	NA
RF2	120	18	800	NA	NA
RF3	120	22	400	NA	NA
RF4	120	22	800	NA	NA
RF5	180	18	400	NA	NA
RF6	180	18	800	NA	NA
RF7	180	22	400	NA	NA
RF8	180	22	800	NA	NA
LB1	120	18	400	Fine	12.6
LB2	120	18	800	Coarse	12.6
LB3	120	22	400	Coarse	12.6
LB4	120	22	800	Fine	12.6
LB5	180	18	400	Coarse	12.6
LB6	180	18	800	Fine	12.6
LB7	180	22	400	Fine	12.6
LB8	180	22	800	Coarse	12.6
HB1	120	18	400	Coarse	24.4
HB2	120	18	800	Fine	24.4
HB3	120	22	400	Fine	24.4
HB4	120	22	800	Coarse	24.4
HB5	180	18	400	Fine	24.4
HB6	180	18	800	Coarse	24.4
HB7	180	22	400	Coarse	24.4
HB8	180	22	800	Fine	24.4

The designs were balanced so that the level of each factor was tested in combination with the levels of the other factors. This ensured an independent and reliable estimation of the effect of each variable (Giesbrecht and Gumpert, 2004). An analysis of variance was performed to evaluate the effect of each factor. This was followed by a multiple comparison test (Least Significant Difference – LSD) to assess whether each level of a parameter was significantly different from the other level (Armitage and Colton, 1998). The full experimental plan applied to the refined wheat flour and the fractional one according to the wheat bran concentration and particle size were analyzed separately. As no significant effect of the bran particle size on the measured properties ($p < 0.05$) was reported, the results are presented depending on the bran concentration. It was not mandatory to replicate samples when using such experimental designs (Giesbrecht and Gumpert, 2004). However the trials with increasing bran concentration extruded at conditions 1, 2, 7 and 8 were repeated 3 times to better investigate the extrusion process and measure the variability of the results. The correlation factors r^2 between the mathematical model and the experimental data are shown in Appendices in Tab. 10.1.

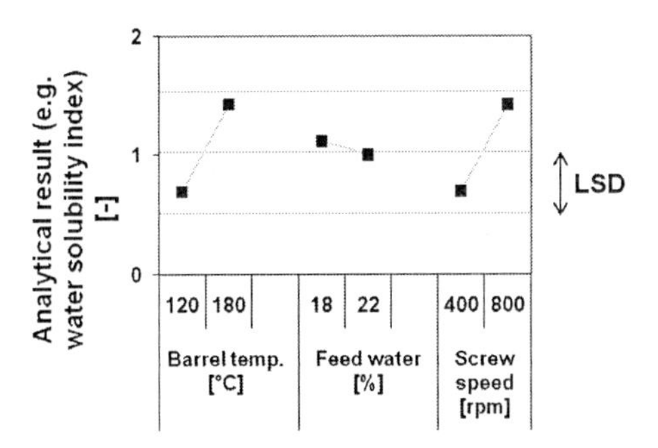

Fig. 3.1: Representation statistical analysis concerning the effect of the process parameters on a measured parameter (the distance between two grid lines represents the least significant difference)

The results of the experimental plan analysis are reported in the manuscript following the example shown in Fig. 3.1. The values of the measured parameter (e.g. water solubility index) are shown depending on the level of the process parameter (low or high level). The values represented in the figures are the average value of the measured parameter when fixed at a given level and when changing the other process parameters level (e.g. when increasing the water content or the barrel temperature). The distance between two grid lines indicates the least significant difference (LSD). The value of LSD means that if increasing the process parameter from the low level to the high one the difference of average values is not greater than the distance between two lines, the effect is not significant. For instance in Fig. 3.1, the effect of changing the level of the barrel temperature and screw speed on the measured parameter would be significant, while it would not be significant when changing the level of the feed water content. The least significant difference (LSD) was calculated for each measured properties indicated above. The least significant differences (LSD) of the measured parameters are shown in the Appendices Tab. 10.1. It was used to estimate if two data are significantly different the one from the other.

Other properties that the ones mentioned above were measured. Nevertheless, they were not measured for all the samples described in the experimental plans. The statistical approach within the experimental plans could not be used. To estimate the significance of the difference for these properties (i.e. melting temperature, glass transition temperature, total, insoluble and soluble dietary fiber content, entrance pressure drop, mechanical properties of the thermomolded bars), a Student T-test was used (two-sided tests).

III.4. Measurement of expansion properties

III.4.1. Bulk expansion properties

The expansion properties were determined by measuring the Volumetric, Sectional and Longitudinal Expansion Indices based on the definitions given by Alvarez-Martinez et al. (1988). The Volumetric Expansion Index (VEI) was defined by equation (3.2):

$$\mathrm{VEI} = \frac{\rho_{\mathrm{m}}}{\rho^*} \frac{(1-\mathrm{W}_{\mathrm{m}})}{(1-\mathrm{W}^*)} \tag{3.2}$$

where ρ and W are the density and the moisture content (wet basis) and indices m and * refer to the melt and extrudate, respectively. The material density, used to estimate the melt density was measured by helium pycnometry (10 replicates, Accupyc 1330, Micrometrics, Verneuil en Halatte, France). The extrudate bulk density ρ^* was measured by beads displacement (three repetitions on 5 pieces).

The relative density D was calculated according to equation (3.3):

$$D == \frac{\rho^*(1-\mathrm{W}^*)}{\rho_s(1-\mathrm{W}_{\mathrm{m}})} = \frac{1}{\mathrm{VEI}} \tag{3.3}$$

And the porosity ε was obtained according to equation (3.4):

$$\varepsilon = (1-D)\times 100 \tag{3.4}$$

The Longitudinal Expansion Index (LEI) is defined as the ratio of the exiting velocity of the extrudate after expansion to the average velocity of the melt in the die and was calculated according to equation (3.5):

$$\mathrm{LEI} = \frac{\pi D_d^2}{4} L_{se}\rho_m \frac{(1-\mathrm{W}_{\mathrm{m}})}{(1-\mathrm{W}^*)} \tag{3.5}$$

D_d is the die diameter and L_{se} is the specific length of the extrudate, defined as the extrudate length, measured with a Vernier caliper. It is given per mass unit of extrudate (average of 10 repeats).

As the Sectional Expansion Index obtained from the measurement of the extrudate diameter with a Vernier caliper was found to lead to a larger experimental error (Della Valle et al. 1997), SEI was preferably calculated according to equation (3.6):

$$\mathrm{SEI} = \frac{\mathrm{VEI}}{\mathrm{LEI}} \tag{3.6}$$

The maximum sectional expansion index at the die exit SEI_{max} was assessed with a digital camera (D40, Sport mode, Nikon, Bern, Switzerland, average of 40 images). The extrudate diameter at maximum expansion D_{max} was measured manually with PhotoFiltre software (v. 6.4.0, Antonio Da Cruz) and related to the die diameter D_d to calculate SEI_{max} according to equation (3.7):

$$SEI_{max} = \left(\frac{D_{max}}{D_d} \right)^2 \quad (3.7)$$

The Shrinkage Ratio (SR) was determined from the sectional (SEI) and maximum sectional expansion index (SEI_{max}) and calculated according to equation (3.8):

$$SR = \frac{SEI_{max} - SEI}{SEI_{max}} \times 100 \quad (3.8)$$

III.4.2. Cellular structure

The characteristics of the cellular structure were analyzed using X-ray tomography. The extruded samples were scanned using a high resolution desktop cone beam X-ray micro-computed tomography system (Scanco µCT 35, Scanco Medical AG, Brütisellen, Switzerland). It consists of a micro-focused sealed X-ray tube operating at a voltage of 55 kV and a current of 145 µA. X-ray shadow images were acquired every 0.18° through a 360° rotation. The reconstruction of the image used a Shepp & Logan filtered back-projection extended to a cone-beam geometry. A voxel size of 6 µm was selected in order to capture the thin cell walls while scanning statistically a significant part of each sample. 3D image analyses were generated using OpenVMS 1-64 software (v.8.3, Hewlett-Packard). A volume of interest (VOI) was selected, segmented and the porosity of the pellet calculated as the ratio of the volume of the cells to the entire VOI. The cell diameter and cell wall thickness distributions were calculated using the method developed by Hildebrand and Rüegsegger (1997). The density N_c of cells was calculated using a component labeling operation. Only the cells larger than the average cell wall thickness were taken into account in the calculation of the cell density to exclude voids comprising only a few voxels and originating from measurement noise. The connectivity index I_{co} was defined as the ratio of the volume of the largest cavity in comparison to the total void volume (Babin et al., 2007). The variability in the measurement of the cell characteristics was less than 5 % (performed on triplicate measurement of a single sample).

III.5. Measurement of material physicochemical properties

III.5.1. Particle size

The particle size distribution of the different flours was assessed using light diffraction (Masterizer 2000, Malvern, Herrenberg, Germany). The measurement was performed in medium-chain triglyceride oil (Cognis Delios V, Firentis A.G., Rheinfelden, Switzerland). The distribution of particle diameter was displayed according to norms ISO 9276. The average volume weighted diameter of the particles $D_{[4,\ 3]}$ was determined from the distribution.

III.5.2. Water Absorption and Solubility Indices

The physicochemical properties of the material were first determined measuring the water solubility index (WSI) and the water absorption index (WAI) according to the method of Anderson et al. (1969). The extruded samples were ground and sieved through a 250 µm mesh. Samples (2.5 g) were dispersed in 30 ml of deionised water, agitated at room temperature (22 °C ± 1 °C) for 30 minutes and centrifuged at 9000 *g* at 25 °C for 15 minutes. The supernatant was dried in an oven overnight at 105 °C and used to calculate the water solubility index (WSI) according to equation (3.9):

$$\text{WSI} = \frac{\text{Supernatant dry weight}}{\text{Sample dry weight}} \times 100 \qquad (3.9)$$

The weight of the sediment was measured to calculate the water absorption index (WAI) according to equation (3.10):

$$\text{WAI} = \frac{\text{Wet sediment weight}}{\text{Sample dry weight}} \qquad (3.10)$$

The water solubility index (WSI) reflects the solubility of the macromolecules present in the material in relation to the total amount of material. When increasing the bran concentration in the sample, the calculation of the water solubility index also takes into account the quantity of dietary fibers in the sample dry weight (see equation 3.9). In order to determine the changes in starch solubility during extrusion and when increasing the bran concentration, the starch water solubility index (SWS) was introduced. It is obtained by dividing the water solubility index by the total starch content in the sample according to equation (3.11):

$$\text{SWS} = \frac{\text{WSI}}{\text{Percentage of starch in the dry sample}} \times 100 \qquad (3.11)$$

III.5.3. Bran properties

III.5.3.1. Total, soluble and insoluble dietary fiber content

The effect of the process conditions on the properties of the dietary fibers was assessed measuring the total, soluble and insoluble dietary fiber content. The amount of total dietary fibers (TDF) was measured according to AOAC 985.29. Only minor modifications to this method were introduced. Starch was gelatinized and partially hydrolyzed with a thermo-stable α-amylase (E-BLAAM, Megazyme Int. Wicklow, Ireland) at 95 °C for 30 min. Proteins were partially hydrolyzed at 60 °C for 30 min with a protease (E-BSPRT, Megazyme Int. Wicklow, Ireland). After adjustment of pH to 4.5, the residual starch was further hydrolyzed to glucose at 60 °C for 30 min with amyloglucosidase (E-AMGDF, Megazyme Int. Wicklow, Ireland). Total dietary fibers were precipitated by adding four volumes of 95 % ethanol, filtered and washed, while insoluble fiber (IDF) content was obtained by filtration after hydrolysis of starch and proteins followed by washing. After drying, protein and ash content were determined. Soluble fiber content (SDF) was obtained by difference of total dietary fibers (TDF) and insoluble dietary fiber content (IDF). Measurements were duplicated.

III.5.3.2. Light microscopy

The effect of extrusion on the morphology (size and dimensions) of bran particles was assessed using light microscopy. The extruded samples were embedded in the resin Technovit 7100 (Kulzer-technik A.G., Wehrheim, Germany). Slices of 5 µm in the longitudinal direction (parallel to the direction of extrusion) were taken using a microtome equipped with a tungsten knife (2055, Leica Geosystems, AG, Heerbrugg, Switzerland). Starch was stained with a 1 % Lugol solution (L6146, Sigma-Aldrich A.G., Buchs, Switzerland), and proteins with a Light green solution (Fluka 62110, Sigma-Aldrich A.G., Buchs, Switzerland). The slices were investigated using a light microscope (Zeiss Axioplan, Carl Zeiss A.G., Feldbach, Switzerland), equipped with a color camera (Axicam MRc5, Carl Zeiss A.G., Feldbach, Switzerland).

III.5.4. Starch supramolecular and molecular structure

Additionally to the water solubility and absorption index indices, the effect of extrusion on the starch physicochemical properties was measured using the analytical tests described in the following subchapters.

III.5.4.1. Pasting properties

Pasting profiles of extruded samples were evaluated using a Rapid Visco Analyzer (RVA-4, Newport Scientific, Jessup, Maryland). Ground extruded samples (< 250 µm, 20 % d.m.

0.1 M $AgNO_3$) were left 15 min at room temperature prior to measurement to allow hydration of the solid material. The samples were hold 1 min at 50 °C, heated to 95 °C at 11 K min^{-1}, held at 95 °C for 3 min and cooled to 50 °C at 6.5 K min^{-1} under stirring at 160 rpm. The peak viscosity η and corresponding time were recorded using the software Thermocline (v. 2.2, Newport Scientific, Jessup, Maryland). Measurements were duplicated.

III.5.4.2. Starch gelatinization temperature

The gelatinization temperature of starch was measured in excess of water (ratio between the sample (d.m.) and water of 1:3, sealed mid-pressure aluminum pan (Mettler Toledo, Greifensee, Switzerland) using a dynamical differential scanning calorimeter (DSC 823e, Mettler Toledo, Greifensee, Switzerland). The samples were heated from -5°C to 160 °C at 5 K min^{-1}. The thermographs were analyzed with Star System® v.9.01 software (Mettler Toledo, Switzerland). The dissociation temperature of starch crystallites was determined at the peak of the transition T_{pk}.

III.5.4.3. Molecular size distribution

Starch molecular size distribution was obtained by gel permeation chromatography. Approximately 200 mg of sample (starting material and ground extruded material) were hydrated in 1 ml of deionized water for 15 min. After adding 10 ml of dimethylsulfoxide the sample was heated in a boiling water bath for 15 min and then left overnight at room temperature (22 °C ± 1 °C) with continuous stirring. The next morning, the samples were reheated in a boiling water bath for 15 min, cooled, centrifuged at 12'000 g for 15 min and filtered on a 0.45 μm filter. The filtrate (200 μl) was injected and eluted through two HR 10/30 columns (GE Healthcare, Glattbrugg, Switzerland) packed with Sephacryl S-1000 (GE Healthcare, Glattbrugg, Switzerland) connected in series with degassed 0.01 M aqueous NaOH at a flow rate of 10 ml h^{-1}, using a precision pump Pharmacia P-500 (GE Healthcare, Glattbrugg, Switzerland). Samples were collected and the carbohydrate content was measured using the phenol-sulfuric acid method (Dubois et al., 1956). The void volume and total elution volume were obtained by injecting waxy wheat starch (S9679, Sigma-Aldrich A.G., Buchs, Switzerland) and glucose (49139, Sigma-Aldrich A.G., Buchs, Switzerland), respectively. Analyzes were duplicated and the results were normalized by dividing the glucose content of each fraction by the total starch content in the sample.

The extruded sample with the highest bran concentration and extruded while introducing the highest specific mechanical energy value was hydrolyzed to glucose using a mixture of thermo-stable α-amylase (E-BLAAM, Megazyme Ltd., Wicklow Ireland) and amyloglucosidase (E-AMGDF, Megazyme Ltd., Wicklow Ireland). The hydrolyzate was run through the permeation column to validate that no fibers were eluted within the total elution volume.

III.6. Measurement of melt rheological properties

III.6.1. Shear viscosity of the melt

Rheology experiments were performed with an adjustable twin-slit rheometer, displayed in Fig. 3.2, with a total slit length L of 250 mm. The slits had a width W of 20 mm and a height H of up to 5 mm. The piston sectional area was L_p (25 mm) x W (20 mm) (Fig. 3.2a). The manufacturing precision of the rheometer parts was 20 µm. The width and the length of the slits were chosen to provide a reasonably wide range of shear rates for extruded food materials. The pistons were placed at the ends of the slits to avoid imposing a high amount of shear and energy dissipation in the region before the measurement zone. This arrangement also avoids early gas bubble nucleation along the slit. An innovative "sliding" system was implemented. It allowed the adjustment of the entire slit heights during the experiments and increasing the range of shear rates and extruder pressure that could be reached (Fig. 3.2b).

Each slit temperature was independently controlled by circulating water inside the metal. The slits were separated by an 8 mm PEEK polymer layer. The temperatures T_1 (measured at a 57.5 mm distance from the second measurement point T_2), T_2, T_3 and T_4 (sensors separated by a distance y_r of 45 mm) along the slit were measured with temperature probes (Pt-100, ROTH+CO. A.G., Oberuzwil, Switzerland). The pressures P_1, P_2 and P_3 (sensors separated by a distance x_r of 45 mm) were measured with piezo-electric pressure sensors (Kistler 4091B02, pressure range from 0 to 10 MPa, Kistler Instrument A.G., Winterthur, Switzerland). The experiments were started with the pistons fully open. Both pistons were then simultaneously advanced to reach a similar melt pressure P_r in the extruded as previously obtained with the extrusion die P_d. Having similar melt pressure in the extruder with the die or the rheometer attached allows having similar thermomechanical history of the ingredients. Slit temperatures were also adjusted to the temperature T_d obtained with the die. Once a stable extruder back pressure P_r was reached, one piston was moved upward, while the other one was moved downward. This allows maintaining the extruder back pressure P_r constant and modulating the flow and shear rate in both slits. When required the temperature of the slits was re-adjusted to be as close as possible to the targeted extruder temperature T_d. The shear rate was also changed by operating the rheometer at different slit heights H. More details on this rheometer configuration are published in Robin et al. (2010b).

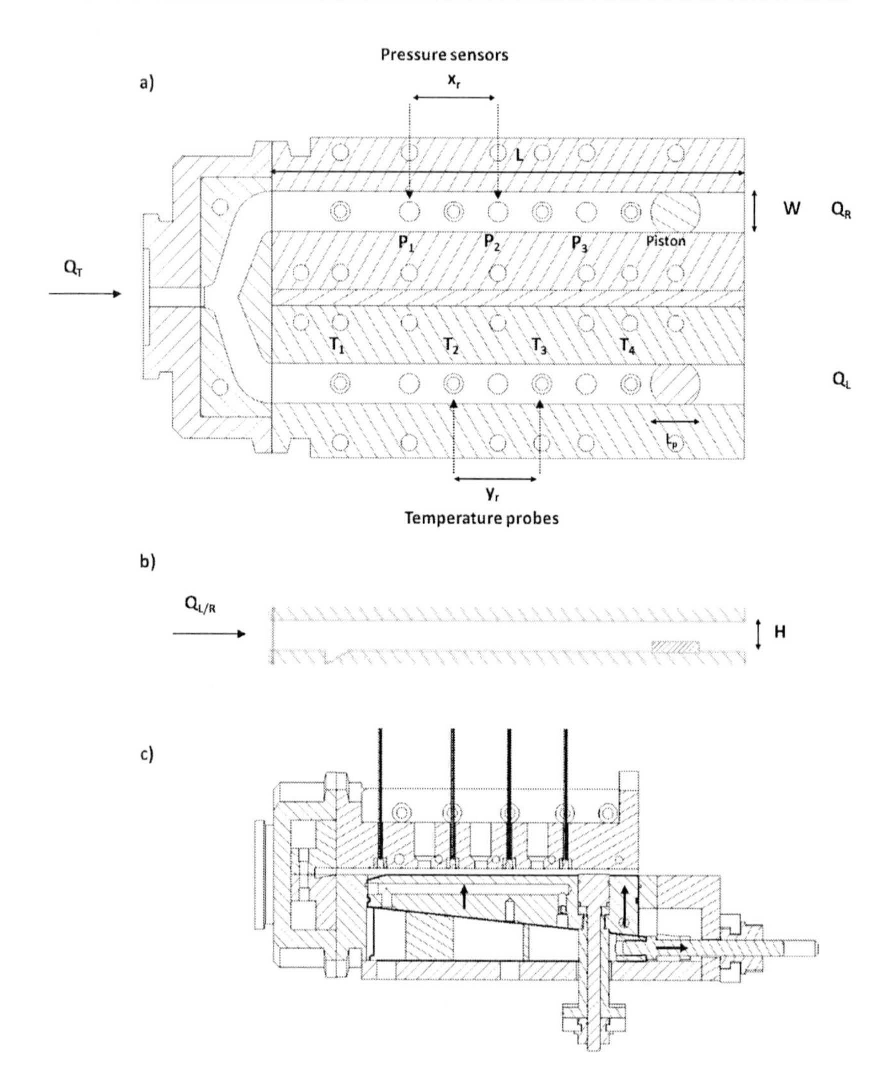

Fig. 3.2: **Schematic drawing of the twin-slit adjustable rheometer (TSAR) (a, cross section with view from above and b, cross section with view from the side) and principle of the rheometer's adjustable slits (c) (H = slit height, h = channel depth under piston valve, L = slit length, L_p = piston length, P = pressure sensory, $Q_{R,\ L}$ = volumetric flow rate in right and left slit, Q_T = total volumetric flow rate, T = temperature probes, W = slit width, x_r = distance between temperature probes, y_r = distance between pressure sensors)**

The apparent shear rate $\dot{\gamma}_a$ was calculated using the geometrical data and the volumetric output rate (Q_L and Q_R), derived using the material's bulk density of the expanded product and according to equation (3.12) (Macosko, 1994).

$$\dot{\gamma}_a = \frac{6Q_{L/R}}{W H^2} \qquad (3.12)$$

On average, four or more shear rates values were applied for each extruder condition. Preliminary experiments showed that an even loss of moisture occurred for the extrudates from both slits, even at the highest temperatures. Therefore the effect of moisture loss on the volumetric output rate was corrected by attributing half of the measured water loss to each slit.

The shear stress τ was calculated according to the pressure drop in the slits ΔP between the pressure sensors separated by a distance x_r using equation (3.13) (Macosko, 1994):

$$\tau = \frac{H}{2(1 + H/W)} \frac{\Delta P}{x_r} \qquad (3.13)$$

The corrected shear rate was calculated from the apparent shear rate $\dot{\gamma}_a$ obtained for a Newtonian fluid and corrected to take into account the flow behavior of the melt using the Rabinowitsch-Weissenberg correction given by equation (3.14) (Macosko, 1994):

$$\dot{\gamma} = \frac{\dot{\gamma}_a}{3}\left(2 + \frac{d\log\dot{\gamma}_a}{d\log\tau}\right) \qquad (3.14)$$

From the shear stress/apparent shear rate relationship, the index n, the shear viscosity $\eta(\dot{\gamma})$ and corrected consistency factor K were obtained assuming that the power law is valid for describing the flow in the slit (Macosko, 1994):

$$\eta(\dot{\gamma}) = K\dot{\gamma}^{n-1} \qquad (3.15)$$

III.6.2. Entrance pressure drop

The effect of the bran concentration on the extensional viscosity of the melt was determined by measuring the entrance pressure as a function of the die length and establishment of the Bagley plot.

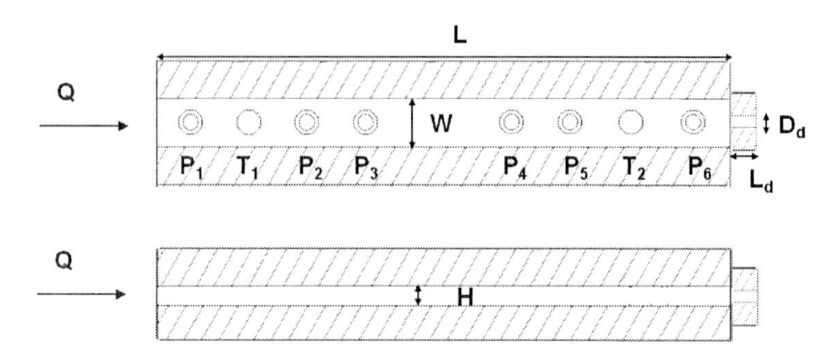

Fig. 3.3: **Design of the single slit rheometer (H = slit height, L = slit length, P_i = Pressure sensors, Q = total volumetric flow rate, T_i = temperature probe and W = slit width) equipped with a cylinder die (D_d = die diameter and L_d = die length)**

For this purpose, a single slit rheometer was attached to a co-rotating double screw extruder (ZSK 26 Mc, Coperion/Werner & Pleiderer, Stuttgart, Germany) and using process conditions described in Chapter III.2. An illustration of the design of the rheometer is shown in Fig. 3.3. Its length L was 370 mm with a width W of 15 mm and a height of 4 mm. Pressure drop in the slit was measured using 6 pressure sensors P_1, P_2, P_3, P_4, P_5 and P_6 (M3 series, Gefran, Daniel Baumann, Germany) positioned along the slit length. The material temperature of the rheometer was set to 150 °C and was measured using two temperature probes T_1 and T_2 (PT 100, ROTH+CO. A.G., Oberuzwil, Switzerland). Three cylindrical dies with a fixed diameter D_d of 3 mm and different length L_d (10 mm, 15 mm and 25 mm) were placed one after the other at the exit of the slit rheometer (Fig. 3.3). The pressure in the slits was recorded over 2 min and the averaged values were used. The Bagley plot was built by plotting the pressure drop ΔP vs. L_d/D_d (with D_d being constant). The pressure drop between the die entrance and atmospheric pressure ΔP_{Ent} was obtained from the linear regression of pressures in the slip from P_2 to P_5 extrapolated to the entrance of the die. The experiments were duplicated.

III.6.3. Establishment of solid state diagrams

The solid state diagram of the material (unprocessed or extruded) with increasing bran concentration was established by measuring the temperature of phase transition (glass transition and melting) depending on the water content. To show the influence of increasing bran concentration on the water distribution between starch and fibers, the sorption isotherms were established.

III.6.3.1. Sorption isotherms

For determining the sorption isotherms at 25 °C, the extruded samples were ground (particle size below 250 μm) and equilibrated at water activities a_w of 0.11, 0.22, 0.33, 0.43, 0.53 and 0.64 by storing them with saturated LiCl, CH_3COOK, $MgCl_2$, K_2CO_3, $Mg(NO_3)_2$ and $NaNO_3$ salt solutions at 25 °C, respectively. Water sorption isotherms were assessed by measuring the water content W (d.b.) corresponding to each water activity a_w. The initial water content in the samples was measured by thermogravimetry (TG-DTA, Q600, TA Instruments, Crawley, United Kingdom). The water gain or loss of the samples during equilibration at a given water activity was monitored during 60 days by weighing the samples. Experimental data were fitted with the Guggenheim-Anderson-de Boer model (Weisser, 1986) according to equation (3.16) and using the Excel solver function.

$$W(a_w) = \frac{W_m C_{GAB} K_{GAB} a_w}{(1 - K_{GAB} a_w)(1 - K_{GAB} a_w + C_{GAB} K_{GAB} a_w)} \qquad (3.16)$$

where C_{GAB} and K_{GAB} are constants and W_m is the theoretical monolayer water content on a dry basis.

III.6.3.2. Melting and glass transition temperatures

DSC thermographs were performed using a computer- and temperature controlled dynamic differential scanning calorimeter (model DSC 823e, Mettler Toledo, Greifensee, Switzerland). 40 mg (d.m.) of samples, previously equilibrated at a given water activity, were placed in a sealed mid-pressure aluminum pan (Mettler Toledo, Greifensee, Switzerland). Changes in hear flow of the tested sample compared to a reference aluminum pan containing water were measured. Determination of the transitions was performed with Star System® v.6.01 software (Mettler Toledo, Switzerland).

For measuring the starch melting temperature, the DSC program comprised a heating ramp from 10 °C to 250 °C at 5 K min^{-1}. The melting temperature of the starch crystallites was measured at the peak of the heat flow corresponding to the melting transition. Experimental points obtained for the melting temperature were fitted with the Flory-Huggins equation (3.17) (Donovan, 1979):

$$\frac{1}{T} - \frac{1}{T^0} = \left(\frac{R}{\Delta H_u}\frac{V_2}{V_1}\right)\left[v_1 - \chi_{12} v_1^2\right] \qquad (3.17)$$

where R is the gas constant, ΔH_u is the enthalpy of fusion of the polymer per repeating unit, V_1 and V_2 are the molar volumes of water and starch, respectively, v_1 and v_2 are the volume fraction of water and starch, T^0 is the melting temperature of the fully dried crystallites and χ_{12} is the Flory-Huggins interaction polymer-diluent parameter.

For measuring the glass transition temperature, a temperature program was selected comprising heating from 5 °C to 140 °C at 5 K min^{-1}, cooling to 5 °C at 10 K min^{-1} and reheating to 150 °C at 5 K min^{-1}. For glass transitions below 50 °C, start and end temperatures of the heating ramps were set to -50 °C and 50 °C, respectively. Glass transition temperature was determined using the second heating ramp to allow structural entropy relaxation and equilibrium. The glass transition $T_{g,\ onset}$ was measured at the onset of the transition to avoid. Experimental points for the glass transition temperature, obtained for each water content, were fitted with the Gordon-Taylor equation (Gordon and Taylor, 1993) according to equation (3.18) and using the Excel solver function.

$$T_{g,onset}(W) = \frac{(100-W)T_{g,m} + k_{G\&T} W T_{g,w}}{(100-W) + k_{G\&T} W} \qquad (3.18)$$

where $T_{g,m}$ and $T_{g,w}$ are the glass transition of the fully dried material and of pure water (-139 °C, as suggested by Orford et al., 1990), respectively and k is the Gordon-Taylor fitting parameter.

The dependency of the melting and glass transition temperatures on water activity were obtained from equations (3.16-3.18) according to equation (3.19) (Palzer, 2004):

$$T_g(a_w) = \frac{(1-K_{GAB}a_w)(1+(C-1)K_{GAB}a_w)T_{g,m} + k_{GAB}W_m CKa_w T_{g,w}}{(1-Ka_w)(1+(C-1)Ka_w) + kW_m CKa_w} \qquad (3.19)$$

III.7. Measurement of mechanical properties of solid foams

The mechanical properties of the cell wall material and of the extruded foams were measured using a three-point bending test on a texture analyzer (TA-HDi, Stable Microsystems, Godalming, UK) equipped with a 50 kg load cell and applying a crosshead speed of 1 mm s^{-1}. The samples were equilibrated prior to testing at a water activity of 0.30 in humidity cabinets (C+ 10/60, CTS A.G., Germany). The mechanical properties of the cell wall material were obtained from thermomolding to eliminate air bubbles and obtain a solid bar of extruded samples. The extruded samples were ground (particle size < 250 µm) and the resulting powder was equilibrated to a water activity of 0.84 using saturated KCl salt solutions at 25 °C. A pressure of 20 MPa and a temperature of 30 K above the glass transition temperature of the sample was applied on the powder placed in a rectangular thermoregulated mold (40 mm long and width W_b of 7 mm). Thus, a solid bar was obtained. The rupture stress σ, maximum strain ε at rupture and elastic modulus E of the obtained bars and of the extruded samples were derived from the maximum force at rupture F and the slope of the linear part of the force vs. crosshead displacement curves (F/d). They were calculated, for the rectangular thermomolded bars (index s) according to equations (3.20, 3.21 and 3.22) (4 repetitions) (Robin et al., 2010a):

$$\sigma_s = F\frac{3L_b}{2e^2W_b} \qquad (3.20)$$

$$\varepsilon_s = \frac{6de}{L_b^{\ 2}} \qquad (3.21)$$

$$E_s = \frac{FL_b^{\ 3}}{4e^3W_bd} \qquad (3.22)$$

e is the average thickness of the bars, W_b their width and L_b the distance between two supports during the bending test ($L_b = 28$ mm).

For the cylindrical extruded samples (index *) the stress at rupture σ^* and the elastic modulus E^* were determined according to equations (3.23, 3.24 and 3.25) (10 repeats) (Alaoui et al., 2008).

$$\sigma^* = \frac{FL_s}{\pi r_e^{\ 3}} \qquad (3.23)$$

$$\varepsilon^* = \frac{3}{4}\frac{dr_e}{L_s^{\ 2}} \qquad (3.24)$$

$$E^* = \frac{4}{3}\frac{F}{d}\frac{L_s^{\ 3}}{\pi r_e^{\ 4}} \qquad (3.25)$$

r_e represents the average radius of the samples and L^* the distance between the two supports ($L^* = 50$ mm). The thickness e of the bars and the radius r_e of the extruded samples were measured with a digital Vernier caliper (average of 10 points).

CHAPTER IV. STARCH PHYSICOCHEMICAL PROPERTIES

IV.1. Introduction and hypothesis

As reported in Chapter II.4.1, the expansion mechanisms of cereal-based extruded products are mainly governed by the physicochemical properties of the plasticized starch matrix. The physicochemical properties of starch are modified during extrusion depending on the process conditions. The hypothesis is that at same process conditions, the addition of bran may modulate the physicochemical properties of the extruded starch and contribute to the observed changes in expansion properties. This may be caused by changes in the melting and glass transition temperatures of starch due to competition for water, modification of the viscosity of the melt and/or protection of starch granules against shear when increasing the bran concentration. To investigate these different hypotheses, starch transformation was assessed depending on the bran concentration using complementary physicochemical tests: water solubility and absorption indices, gelatinization temperature of starch, starch pasting properties and molecular size distribution, glass transition and melting temperatures as well as sorption isotherms. To investigate the effect of extrusion on fiber properties, the total, soluble and insoluble dietary fiber contents were measured. Changes in the physicochemical properties of starch and dietary fibers as a result of the process conditions and bran concentration are reported and discussed in this Chapter.

IV.2. Effect of bran and extrusion conditions on physicochemical properties of starch

IV.2.1. Water solubility and absorption indices

The water solubility and absorption indices of samples with increasing bran concentration which are extruded at different conditions were measured and are depicted in Fig. 4.1. Negative trends were shown between the water absorption (WAI) and solubility indices (WSI). According to Smith (1992), such a trend indicates the presence of both totally damaged starch granules and degradated starch molecules (see in Chapter II.4.1).

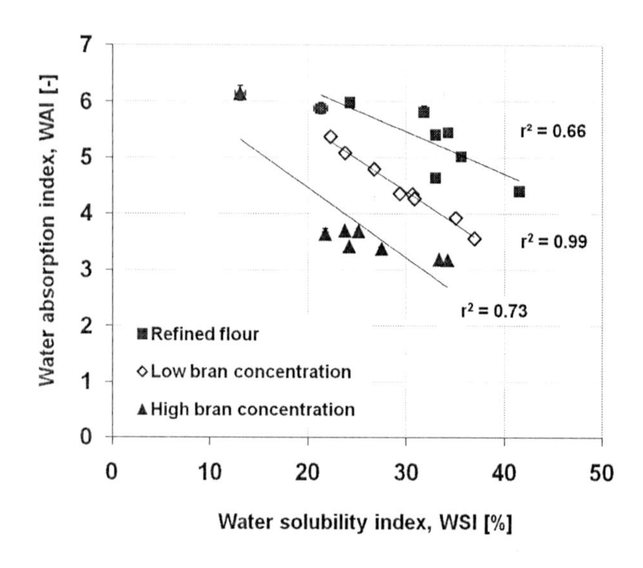

Fig. 4.1: **Water absorption index (WAI) vs. water solubility index (WSI) of samples with increasing bran concentration and extruded under different conditions**

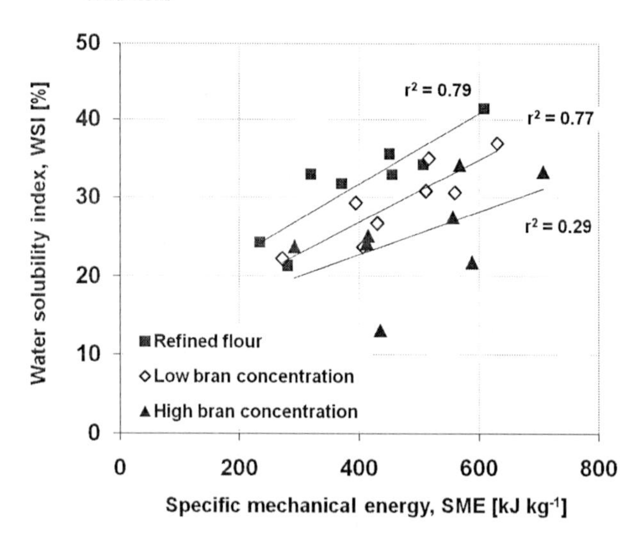

Fig. 4.2: **Water sorption index (WSI) vs. specific mechanical energy (SME) of samples with increasing bran concentration and extruded under different conditions**

The values of water solubility index (WSI) values could be positively correlated with the specific mechanical energy (SME) for the extruded refined flour (RF) and low bran concentrations (LB) as shown in Fig. 4.2 (correlation factor, r^2 = 0.79 and 0.77, respectively). The water solubility and absorption indices were nevertheless weakly correlated for the sample with the highest bran concentration (HB) (r^2 = 0.29). Both results were in agreement to previous reports on extruded cereals low in fibers in the elevated specific mechanical energy range (Schuchmann and Danner, 2000; Robin et al., 2010a). The authors explained it by a degradation of starch macromolecules, mainly amylopectins.

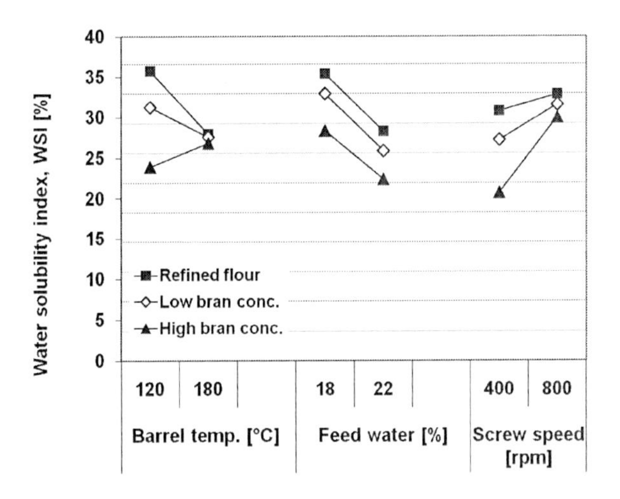

Fig. 4.3: **Effect of process conditions on water solubility index (WSI) of extruded samples with increasing bran content (the distance between two grid lines indicated the least significant difference)**

As shown in Fig. 4.3, the increase in feed water content significantly decreased the water solubility index (WSI) for all samples, regardless of the bran concentration. This decrease in water solubility index with water content was mainly attributed to a decrease in shear viscosity in the extruder (see melt rheology in Chapter V in Fig. 5.9b). The decrease in water solubility index was likely due to a lower degradation of amylopectin during extrusion. The increase in barrel temperature also significantly decreased the water solubility index (WSI) values for the refined wheat (RF) and low bran (LB) concentration samples (Fig. 4.3). This decrease in water solubility index with barrel temperature can be associated to a decrease in the melt viscosity with temperature (see melt rheology in Chapter V in Fig. 5.9b). For the high bran concentration recipe, the difference in average water solubility index between the low and high barrel temperature was lower than the least significant difference (represented by the distance between two grid lines in Fig. 4.3).

This means that effect of the barrel temperature on the water solubility of the high bran (HB) concentration samples was not significant. For this recipe, an opposite trend with temperature could nevertheless be observed (Fig. 4.3). Interacting effect between bran particle and starch might explain this result. Increasing the screw speed significantly decreases the shear viscosity of the shear thinning melt in the extruder (see melt rheology in Chapter V in Fig. 5.9b). Nevertheless, it also increases the shear stress transmitted to the starch molecules, resulting in an increased starch transformation extent and a significant increase in water solubility index, indicating increased amylopectin degradation (Fig. 4.3) (e.g. Barrès et al., 1990).

IV.2.2. Estimated starch solubility

In order to investigate the change in starch solubility according to the bran content, the water solubility index has to be normalized by the starch content in the material. For this the starch water solubility value was introduced (see in Chapter III, equation 3.11). This value can be used as the fiber and protein solubility was low and had no significant effect on the water solubility index depending on the process conditions. Indeed as it can be seen later in this Chapter in Table 4.2 that no significant changes in soluble dietary fiber content (SDF) could be reported after extrusion, irrespective of the process conditions and bran concentration. Additionally, protein solubility in water was shown to be low and reduced after extrusion (Falcon and Phillips, 1988; Yoshii et al., 1990; Lei and Lee, 1996).

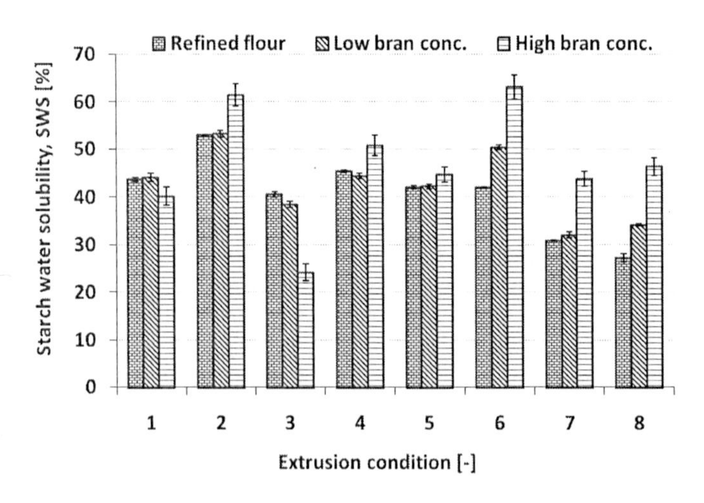

Fig. 4.4: Starch water solubility (SWS) when increasing the bran concentration at same extrusion conditions (see in Chapter III in Tab. 3.2 for the extrusion conditions)

The values of starch water solubility according to the bran concentration and process conditions are shown in Fig. 4.4. The results showed that the starch water solubility values of the extruded samples without and with added bran at the intermediate concentration were close, except for conditions 6 (180 °C, 18 % feed water content, 800 rpm) and 8 (180 °C, 22 % feed water content, 800 rpm). The samples with the highest bran concentration had higher starch water solubility values than bran-containing samples, except for condition 1 (120 °C, 18 % feed water content, 400 rpm) and 3 (120 °C, 22 % feed water content, 400 rpm). These differences in starch water solubility were process conditions-dependant. The highest differences between the two bran-containing samples were obtained at the highest specific mechanical energy input (e.g. significant difference of 13 % at 292 kJ kg^{-1} and 21 % at 567 kJ kg^{-1}).

IV.2.3. Rapid Visco Analysis and pasting profiles

Fig. 4.5 shows the RVA analysis results of the raw material and of the extruded material at the two extreme values of specific mechanical energy (SME) input. At condition 2, (120 °C, 18 % feed water content, 800 rpm) the highest value of SME is obtained. At condition 7 (180 °C, 22 % feed water content, 400 rpm), the lowest value of SME input is obtained.

For the raw material (not extruded), a peak of viscosity, corresponding to the maximum swelling of non disrupted starch granules, can be observed (Fig. 4.5). The peak of viscosity measured for the refined wheat flour (RF, η_{RVA} = 2100 mPa s) was only slightly reduced when increasing the bran concentration at the low level (LB, η_{RVA} = 2000 mPa s) (Fig. 4.5). The peak of viscosity was drastically decreased when further increasing the bran to the higher concentration (HB, η_{RVA} = 860 mPa s) (Fig. 4.5). This decrease in peak viscosity is likely due to the change in composition, with mainly a decrease in starch content when increasing the bran concentration. After extrusion, a peak of viscosity in the RVA profile could still be observed for all samples (Fig. 4.5). However it occurred at a shorter measurement time than for the unprocessed material (Fig. 4.5).

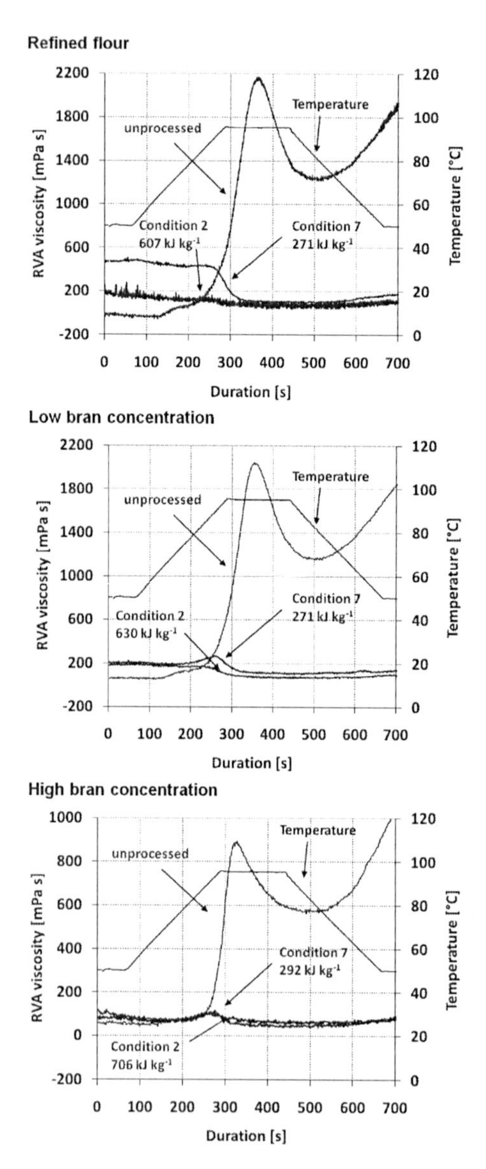

Fig. 4.5: **Pasting profiles of refined flour, low bran concentration and high bran concentration samples unprocessed or extrusion at condition 2 (120 °C, water content in the feed of 18 % and 800 rpm) or at condition 7 (180 °C, water content in the feed of 22 % and 400 rpm) (specific mechanical energy (SME) is indicated next to the sample name)**

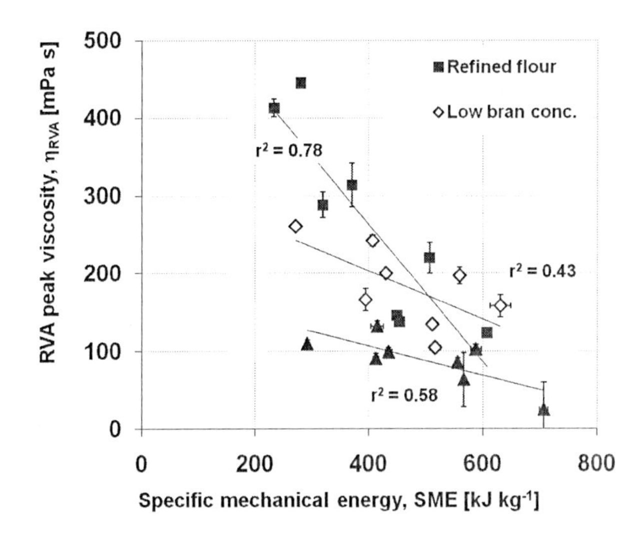

Fig. 4.6: **RVA peak viscosity (η_{RVA}) vs. specific mechanical energy (SME) for refined flour, low bran concentration and high bran concentration samples extruded at different conditions**

The RVA peak viscosity of the extruded refined flour samples and the specific mechanical energy (SME) input appeared to be negatively correlated Fig. 4.6 (correlation factor r^2 = 0.78). The correlation between the RVA peak viscosity and the specific mechanical energy was much poorer for the low and high bran concentration (r^2 = 0.43 and 0.58, respectively). Swelling of starch granules is mainly attributed to changes in amylopectin structure and crystallinity (Tester and Morrison, 1990). According to DSC data (not shown here), the endothermic peak of starch around 66 °C, corresponding to the gelatinization temperature of the unprocessed wheat flour, was not detected any longer after extrusion. The absence of the endothermic peak around 66 °C indicates a total loss of the starch gelatinization peak, irrespective of the bran concentration and process conditions. Therefore, after extrusion, the samples are in the amorphous state. According to the pasting profile of extruded samples and the presence of a peak viscosity, starch structures that could swell in hot water remained after extrusion. The presence of such structures after cooking extrusion was rarely reported in the literature. Hagenimana et al. (2006) mentioned extruded rice-based material still able to swell under heat and an excess of water. They also reported a lack of DSC gelatinization peak, characteristics of intact starch granules.

IV.2.4. Distribution of starch molecular size

To analyse the effect of process conditions and bran concentration on the molecular size profile of starch, gel permeation chromatography was used.

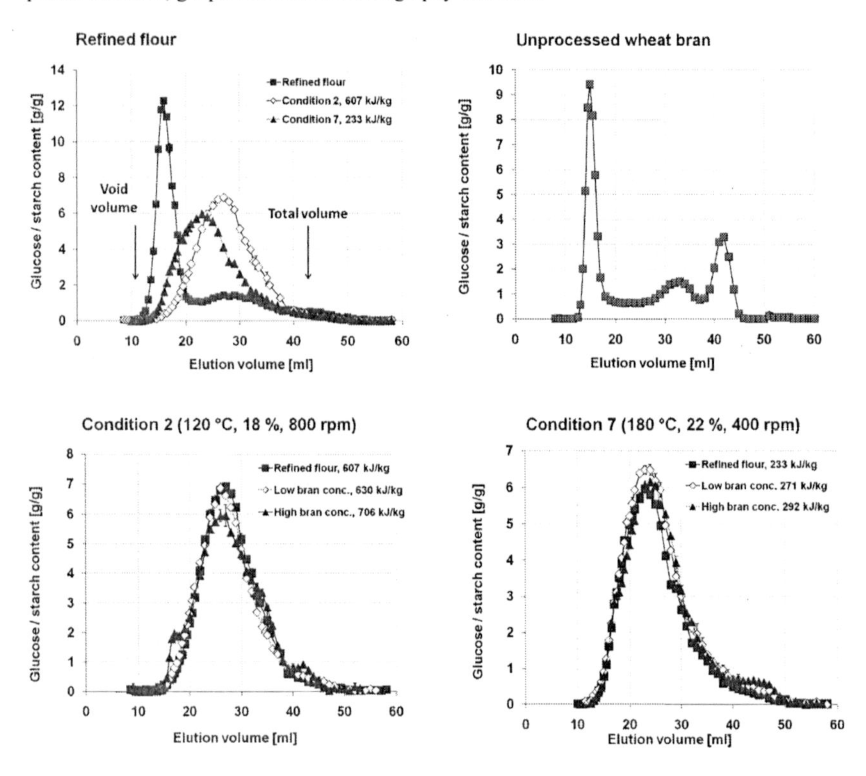

Fig. 4.7: Molecular size distribution of unprocessed refined wheat flour and wheat bran and extruded samples at condition 2 and 7 (the specific mechanical energy (SME) is indicated next to the sample name)

Fig. 4.7a shows the results of the unprocessed wheat flour and the extruded samples at two extreme conditions of specific mechanical energy (conditions 2 (120 °C, 18 % feed water content, 800 rpm): highest value and condition 7 (180 °C, 22 % feed water content, 400 rpm: lowest value). The unprocessed material showed two elution peaks: a high molecular size peak, corresponding to amylopectin and one with a lower molecular size corresponding to amylose. The molecular size distribution of starch was significantly modulated by the extrusion process (Fig. 4.7). Irrespective of the composition, a unimodal distribution was observed after extrusion. This peak was shifted to lower molecular sizes with increasing specific mechanical energy input. At lowest specific mechanical energy input (condition 7), the distribution overlapped the amylopectin and amylose peaks of the unprocessed refined flour (Fig. 4.7). At the highest specific mechanical energy input (condition 2) the amylopectin peak almost disappeared, resulting in a distribution closer to that of pure amylose (Fig. 4.7). This supports the water absorption and solubility indices results (see § IV.2.1), tracing changes in extruded product behavior back to the degradation of amylopectin branches, as also described by (Davidson et al., 1984; Kingler, et al., 1986; Politz et al., 1994; Schuchmann and Danner, 2001, Brümmer et al., 2002). Some authors also reported a change in both amylose and amylopectin molecular size ditribution (Colonna et al., 1984).

The molecular size distribution of starch at different bran concentration but at same process conditions is compared in Fig. 4.7. At the lowest specific mechanical energy (condition 7) and highest specific mechanical energy (condition 2), the elution profiles were very close, irrespective of the bran content. Nevertheless, at the highest specific mechanical energy (condition 2) a slight difference in molecular size distribution could be observed for the highest bran content showing a shoulder peak close to the void volume. This could indicate remaining amylopectin chains (Fig. 4.7). The close elution profile of the refined wheat flour and low bran concentration recipes confirms that the degree of starch transformation may be similar at no and low bran concentration. This is also confirmed by close starch water solubility values.

The higher percentage of starch soluble starch (SWS) for the highest bran content samples compared to the low or no bran concentration samples at low specific mechanical energy input (condition 7: SWS = 44 % vs. 32 % and 31 %, respectively) and high specific mechanical energy input (condition 2: SWS: 61 % vs. 53 % and 53 %, respectively) (see Fig. 4.4) might then be due to the following reasons:

i) The increased disruption of the previously mentioned amorphous structures which were shown to be able to hydrate and swell in hot water during the RVA measurement. This might lead to the release of some soluble starch molecules,

ii) The generation of lower molecular weight carbohydrates in extrusion. On the gel permeation chromatographs of the extruded samples (Fig. 4.7), a peak close to the total elution volume could be observed. Its amplitude was increased with the bran concentration. This peak could be attributed to the bran fraction as shown on its gel permeation chromatograph (Fig. 4.7). The generation of low molecular weight starch fractions during extrusion (likely resulting from amylopectin chains degradation) may overlap with this peak. It is however challenging to quantify them on the chromatographs.

IV.2.5. Solid state diagrams

Starch and wheat fiber may have different hygrocapacity. At constant water content, the distribution of water between the different matrix phases in the extruder may be changed depending on the bran concentration. Depending on the water distribution and the level of water available in the difference phases, the transformation of the macromolecules (e.g. starch and fibers) may vary. During extrusion this distribution in water may be further changed depending on the starch and fiber physicochemical state. This change in water distribution between the ingredients may affect the expansion properties at the die exit. In order to explain the transformation of starch during extrusion, sorption isotherms, starch melting and glass transition temperatures of the recipe with increasing bran concentration were investigated for both the unprocessed and extruded material.

IV.2.5.1. Sorption isotherms

IV.2.5.1.1. Unprocessed material

The sorption isotherms of the unprocessed material with increasing bran concentration are shown in Fig. 4.8.

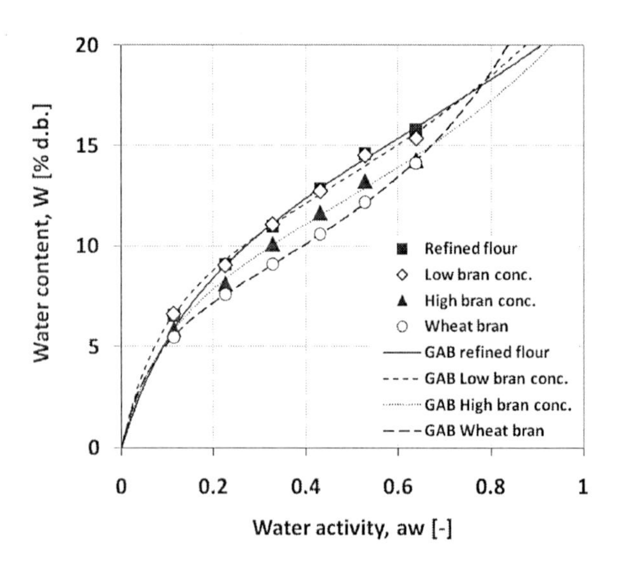

Fig. 4.8: Sorption isotherms of unprocessed refined flour, low bran concentration, high bran concentration and wheat bran (GAB representation)

The fitting parameters of the GAB model are shown in Appendices in Tab. 10.3. The hygrocapacity of the unprocessed bran was significantly lower than the one of the unprocessed refined flour. The main difference in composition between these two ingredients is an increase in dietary fiber and a decrease in starch content. Therefore the lower hygrocapacity of wheat bran can be attributed to the dietary fibers. Nevertheless, as shown in Fig. 4.8, there was no significant difference in the sorption isotherms between the refined flour and the low bran concentration. Therefore, the amount of fibers in this sample was not sufficient enough to induce a change in the hygrocapacity of the matrix.

IV.2.5.1.2. Extruded samples

The sorption isotherm of the samples extruded at 120 °C, 18 % feed water content, 400 rpm (condition 1) and at 180 °C, 22 % feed water content, 800 rpm (condition 8) are shown with the GAB representation (equation 3.16) in Fig. 4.9. These samples were generated with two level of specific mechanical energy input.

The fitting parameters of the GAB model can be found in Appendices in Tab. 10.4. The sorption isotherm of the extruded samples was different from the one of the unprocessed material. Especially at higher water concentrations a significant difference was observed. An example of this comparison between the unprocessed refined flour and the refined flour extruded at condition 1 is given in Appendices in Fig. 10.2. Within the studied range of water activities, the hygrocapacity of the refined flour was decreased after extrusion. A similar observation was made for the bran-containing samples. These results are consistent with the findings for potato starch after hydrothermal treatment (Stute, 1991) or results obtained for gelatinized cassava (Gevaudan et al., 1989). Change in granular structure, crystallinity and molecular weight, reported in Chapter IV.2.3 may be responsible for the lower hygrocapacity of the material after extrusion.

As observed for the unprocessed material, the hygrocapacity of the processed samples was reduced when increasing the bran concentration. The refined flour sample extruded at condition 8 appeared to have a higher hygrocapacity compared to the refined flour sample extruded at condition 1 at the highest water activities (Fig. 4.9). This may be linked to a lower molecular weight of starch as the sample extruded at condition 1 was extruded at higher specific mechanical energy than sample 8 (SME = 506 kJ kg^{-1} vs. 280 kJ kg^{-1}). This led to a higher water solubility index for samples 1 compared to sample 8 (WSI = 34.2 % vs. 21.3 %). A similar increase in hygrocapacity was reported for carbohydrate matrices when decreasing their molecular weight (see e.g. Ubbink et al., 2007). On the opposite, there were only slight differences between the bran-containing samples extruded at condition 1 or at condition 8. This might be linked to the dilution of starch or closer starch physicochemical properties as shown by their closer WSI at high bran concentration (WSI = 21.8 % and 25.1 % respectively at condition 1 and 8).

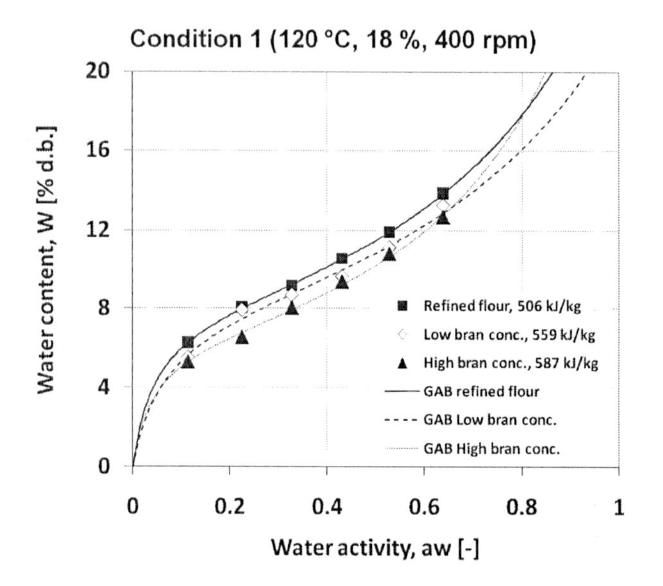

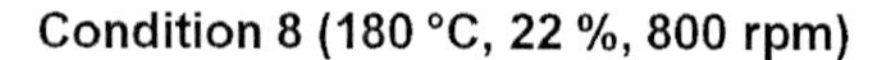

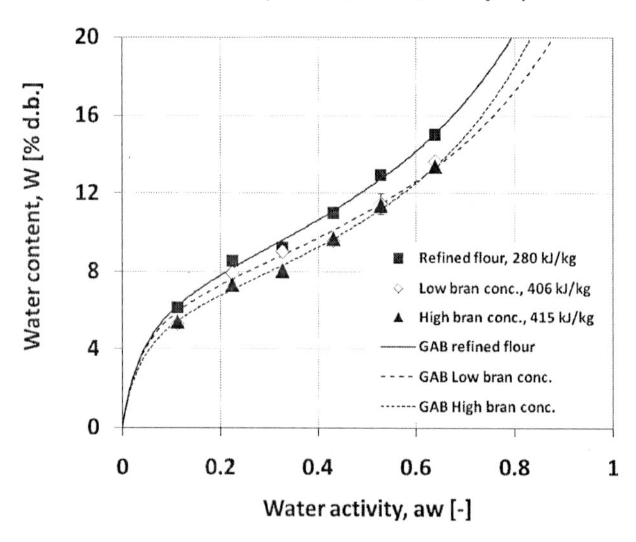

Fig. 4.9: **Sorption isotherms of extruded samples with increasing bran concentration extruded at condition 1 and at condition 8 (specific mechanical energy (SME) is indicated next to the sample name)**

IV.2.5.2. Melting temperature

Melting temperature is associated with an endothermic heat transmission peak. When increasing the bran concentration, a change in matrix composition and hygrocapacity is observed (see Chapter IV.2.5.1). The melting temperature of starch changes. The DSC thermographs for different water volume fractions v_1 are shown in Fig. 4.10 for different bran concentrations in comparison to pure wheat bran. For the extruded refined flour, at $v_1 = 0.09$, a large endothermic transition with several overlapping peaks can be observed at a temperature of about 207 °C (Fig. 4.10). The temperature at the peak shifts towards lower values when increasing the water content. At the highest water content ($v_1 = 0.20$), this endothermic peak is obtained at about 161 °C. It becomes broader with water, indicating more heterogeneous crystals. The temperature of the endothermic peak which occurs at $v_1 = 0.20$ appears to be close the one reported by Jang and Pyun (1996) for wheat starch and a similar water content. The authors reported a peak at about 175 °C at 9.5 % water content close to the value found in this study at 9.92 % water content (T_{pk} = 180 °C). This peak was attributed to the combination of two peaks: one corresponding to the melting of starch crystallites (Burt and Russel, 1983) and a second one to the melting of the remaining crystallites or melting of amylopectin crystallites (Shogren, 1992). An additional shoulder can be detected at $v_1 = 0.12$ at about 205 °C. This indicates the presence of different types of complexes (Fig. 4.10). Two or more peaks with large enthalpies are found at the highest water volume fractions at about 200 °C. Endothermic peaks close or above 200 °C were also reported by Jang and Pyun (1996) for wheat starch in a similar range of water contents.

When adding bran at a low fiber concentration, similar observation could be made (Fig. 4.10). Nevertheless, the temperature, at which the endothermic melting peak is found, is slightly lower. It could hardly be detected at $v_1 = 0.18$ and 0.19. This may be explained by the reduced starch content and/or overlap with other transitions associated with the starch or with the other flour components. For high bran content and pure wheat bran, the endothermic peak with the largest enthalpy value decreases up to a water volume fraction v_1 of 0.15 (Fig. 4.10). It is not detectable any longer for higher water volume fractions (Fig. 4.10). This endothermic transition may be attributed to the melting of starch crystallites. At same volume fraction of water, the peak temperature of this endothermic transition was lower for the high bran content and wheat bran material than for the refined flour or the low bran concentration. This may indicate an effect of the bran fibers on the melting of starch crystallites. A similar effect of oat fibers on the melting of starch crystallites was reported by Nuñez et al. (2009).

The temperature of the endothermic transition was determined at the peak (indicated by T_{pk} in Fig. 4.10) for the different water volume fractions v_1. Temperature at the peak was chosen due to the difficulty of estimating the end of the transition. The standard deviations for T_{pk} did not exceed 4K. The Flory-Huggins equation (Chapter III, equation 3.17) was used to fit the experimental points. This fitting equation is better suited for water diluted starch systems. Nevertheless, it was applied with fair agreements in several studies to more complex systems such as flours (e.g. Núñez et al., 2009; Champenois et al., 1995). The experimental T_{pk} values of all the material with increasing bran concentration were plotted against the water volume fraction.

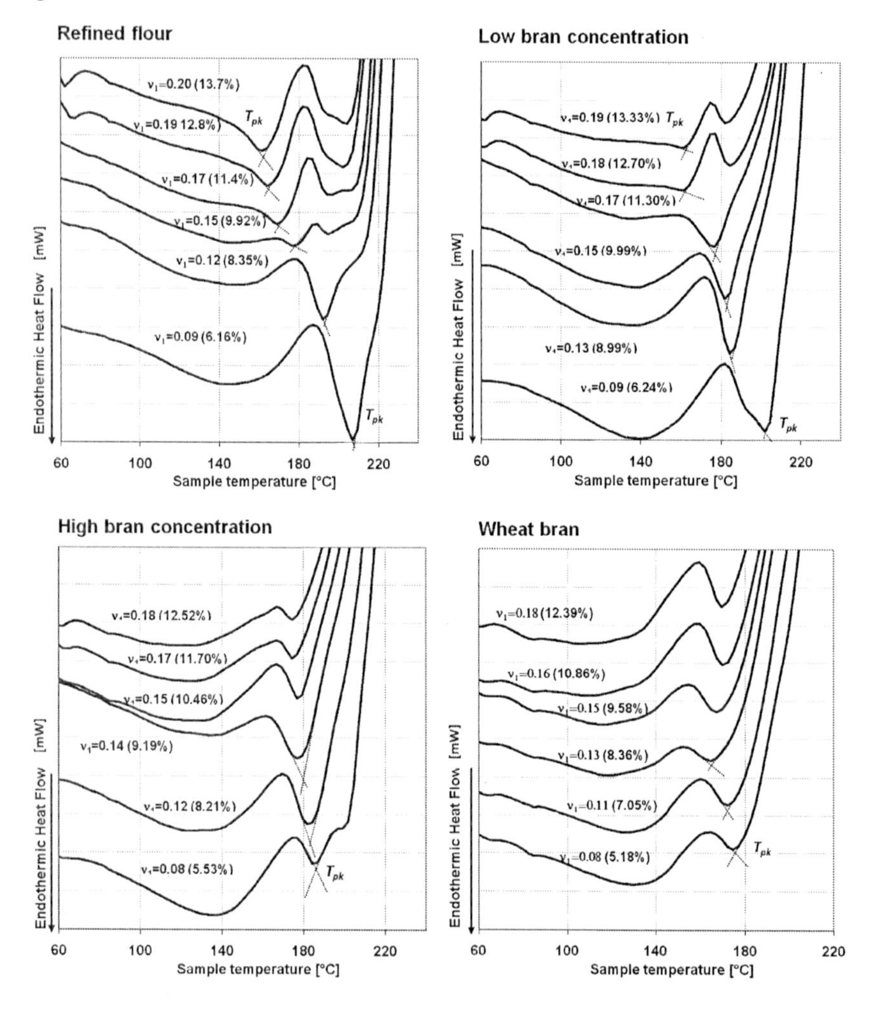

Fig. 4.10: DSC analysis: Endothermic heat flow of unprocessed material with increasing bran concentration depending on the water volume fraction v_1 (Corresponding value of water content is given between parenthesis) (T_{pk} is the melting temperature at the peak of the transition)

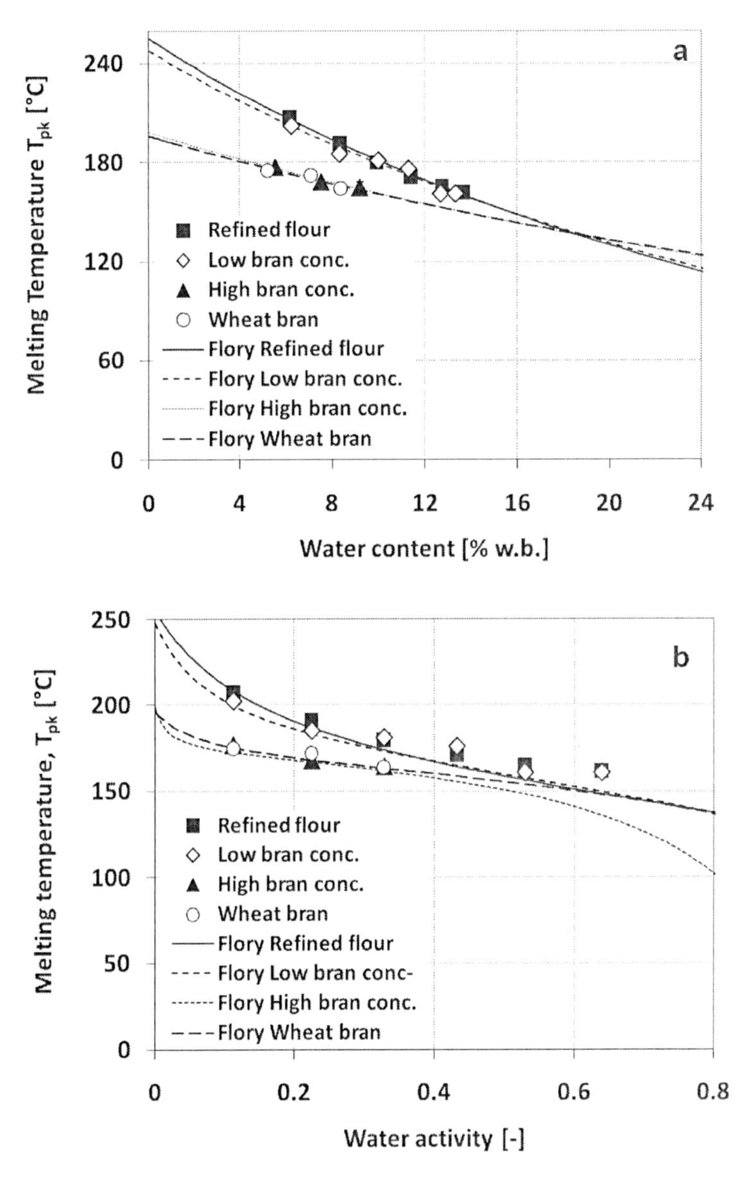

Fig. 4.11: Melting temperature at the peak T_{pk} of unprocessed material as a function of water content W (a) and water activity a_w (b) (Flory-Huggins representation)

Even though only three points could be extracted from the DSC thermographs for the high bran concentration and wheat bran samples (Fig. 4.11a), these plots were established. As expected from literature, T_{pk} decreases with the water content. The relationship was nearly linear and within the range of water volume fraction ($< v_1 = 0.7$) in which an approximation is accepted (Jang and Pyun, 1996). Therefore, the Flory-Huggins parameter χ_{12} was chosen to 0. The melting temperature of the dry starch crystallite T^0_{pk} was extrapolated from the fit. These values of T^0_{pk} were close for refined flour (238 °C) or low bran concentration (236 °C). The value T^0_{pk} while they were lower for high bran concentration (198 °C) and wheat bran (186 °C). Fitting parameters of the Flory-Huggins model are displayed in Appendices in Tab. 10.3.

Although the hygrocapacity of refined flour and flour at low bran concentration were close, it is significantly reduced for high bran concentration and wheat bran (see sorption isotherms in Fig. 4.8). At same water activity, the melting temperature T_{pk} appeared to be significantly reduced when increasing the bran concentration to its highest levels as shown in Fig. 4.11b. This shows a redistribution of the water among the different phases. Since the wheat fibers are less hydrophilic that starch, this indicates that more water was available for the starch phase, decreasing its melting temperature. At the same water content, the increase in free water available for the starch phase leads to a decrease in melting temperature of starch (Fig. 4.11a).

IV.2.5.3. Glass transition temperature

IV.2.5.3.1. Unprocessed material

The change in matrix composition and hygrocapacity induced by the addition of bran may also change the glass transition of starch. The glass transition temperature was measured at the onset of the transition $T_{g,\ onset}$ as shown in Fig. 4.12. The transition was not a sharp transition as earlier observed by some authors for cereals (Kaletunç and Breslauer, 1996).

The glass transition temperature $T_{g,\ onset}$ of the unprocessed material depending on the water content is reported in Fig. 4.13a. The standard deviations of $T_{g,\ onset}$ did not exceed 4 K. This transition can be mainly attributed to starch (Cuq et al., 2003). As expected the $T_{g\ onset}$ of the different samples decreased with the increase in water content, regardless of the bran concentration. The $T_{g\ onset}$ values of the unprocessed refined flour samples were lower than those reported by Kaletunç and Breslauer (1996) for wheat flour and by Zeleznak and Hoseney (1997) for native wheat starch. Nevertheless in these studies the glass transition temperature was measured at the inflexion point of the transition

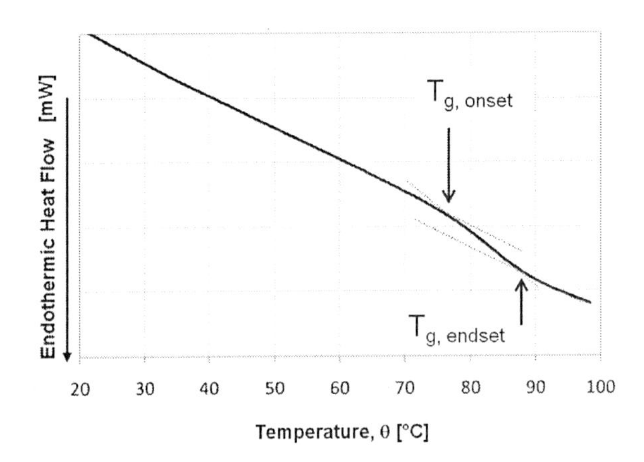

Fig. 4.12: Differential scanning calorimetry thermograph of native refined wheat flour equilibrated at 0.33 water activity (water content of 9.92 %) (second scan)

The experimental data were fitted with the Gordon-Taylor model (equation 3.18) (Gordon and Taylor, 1993). The fitting parameters of the model can be found in Appendices in Tab. 10.3. From extrapolation the glass transition temperature of the dry material $T_{g,m}$ was calculated. With increasing bran concentration the melting temperature $T_{g,m}$ decreased from 194 °C to 105 °C. This was also true for increasing water contents, where a linear relation between bran content and decrease in glass transition temperature $T_{g,\ onset}$ was found. The extrapolated value of glass transition temperature $T_{g,m}$ for the refined flour was lower than the value measured on a fully dried wheat flour (about 167 °C, measured at the inflexion point) (Kaletunç and Breslauer (1996).

A similar observation could be made for the extruded and unprocessed material; at the same water activity, the glass transition temperature was significantly reduced when increasing the bran concentration (Fig. 4.13b). This indicates a shift in the water distribution between the different phases of the matrix. This also indicated that more free water available for the starch phase. At same water content, this higher amount of the free water resulting from the increase in bran content leads to a reduction in glass transition temperature of starch (Fig. 4.13a). Effect of other cereal components, such as proteins (see Cuq et al., 2003) on the glass transition temperature may also interfere with the fibers effect.

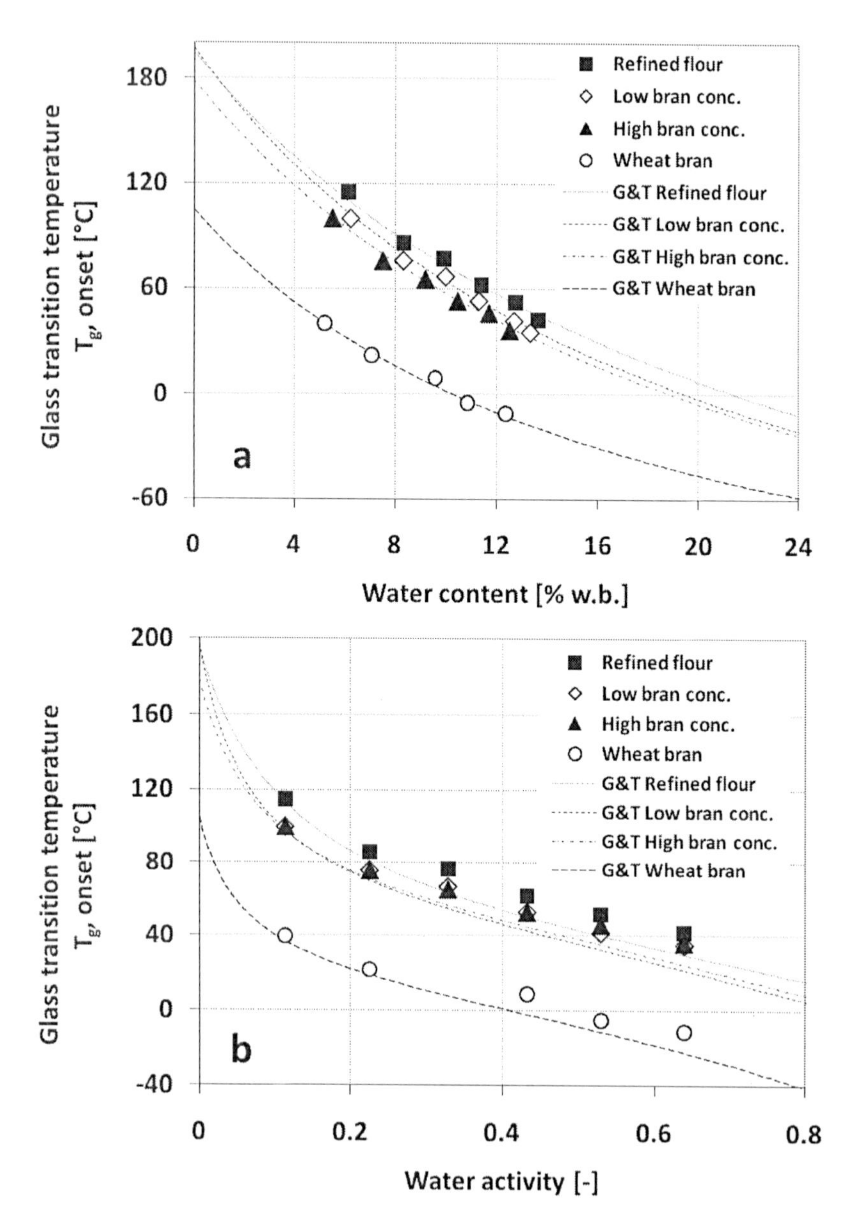

Fig. 4.13: Glass transition temperature (onset $T_{g,\,onset}$) as a function of water content W (a) and water activity a_w (b)

IV.2.5.3.2. Extruded samples

As for the unprocessed material, the reported differences in hygrocapacity when increasing the bran concentration (see Fig. 4.9) may lead to significant differences in the glass transition temperature of starch.

As observed for the unprocessed material, increasing the water content decreases the glass transition temperature $T_{g,\ onset}$ (Fig. 4.14). Increasing the bran concentration to the highest level significantly reduced the glass transition temperature of starch measured at the onset of the transition $T_{g,\ onset}$, regardless of the water content and extrusion condition as shown in Fig. 4.14 for both conditions (the standard deviation of $T_{g,\ onset}$ was below 4 K). Using the Gordon-Taylor model (see equation 3.18) extrapolated to a water content of zero, the glass transition of the dried material $T_{g,m}$ can be estimated. The glass transition temperature $T_{g,m}$ was 187 °C, 148 °C and 135 °C respectively for the refined flour, low bran and high bran concentration material extruded at condition 1. At condition 8 $T_{g,m}$ was 161 °C, 140 °C and 154 °C respectively for the refined flour, low bran and high bran concentrations (fitting parameters of the G&T models are shown in Appendices Tab. 10.4). The $T_{g,m}$ values of the extruded samples were lower than those of the unprocessed material which were respectively for the refined wheat flour, low bran and high bran concentrations 194 °C, 197°C and 178 °C.

As reported in for the unprocessed material, for all bran concentrations, the glass transition temperature $T_{g,\ onset}$ was reduced with the water activity as shown in Fig. 4.15. This reduction in glass transition temperature was higher at condition 1 (120 °C, 18 %, 400 rpm) and only significant at low water activities (0.22 and 0.33). At same water activity the decrease in glass transition when increasing the bran concentration was lower for the extruded samples than for the unprocessed material (see Fig. 4.13b and 4.15). The decrease in glass transition temperature of the extruded samples with added bran may indicate a change in the distribution of water inside the matrix. This may results in more free available water for the starch phase, decreasing its glass transition temperature (see Fig. 4.14). Nevertheless, the distribution of water within the starch and fiber phases appears to be different for the unprocessed and processed samples.

Condition 1 (120 °C, 18 % and 400 rpm)

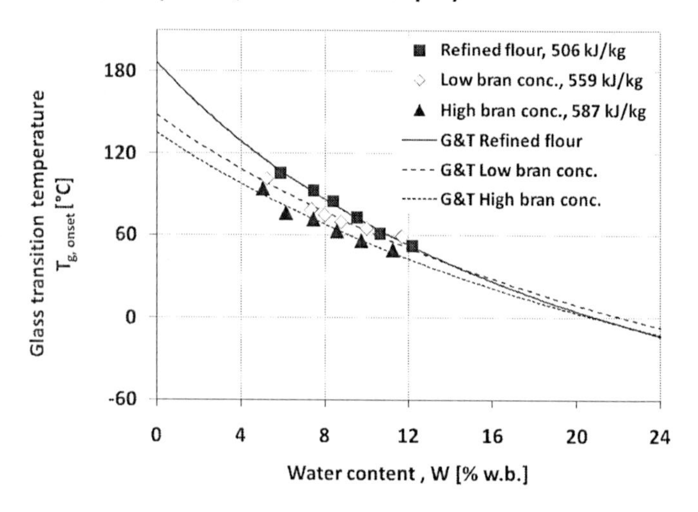

Condition 8 (180 °C, 22 % and 800 rpm)

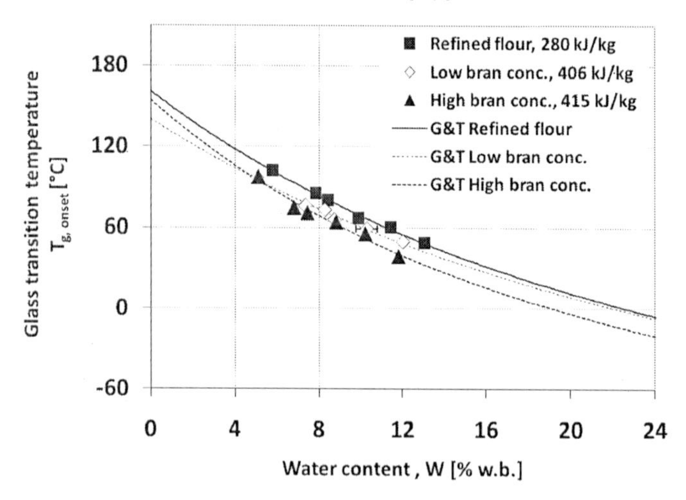

Fig. 4.14: **Glass transition temperature $T_{g,\ onset}$ depending on the water content W of samples with increasing bran concentration extruded at condition 1 and condition 8 (specific mechanical energy (SME) is indicated next to the sample name)**

Condition 1 (120 °C, 18 % and 400 rpm)

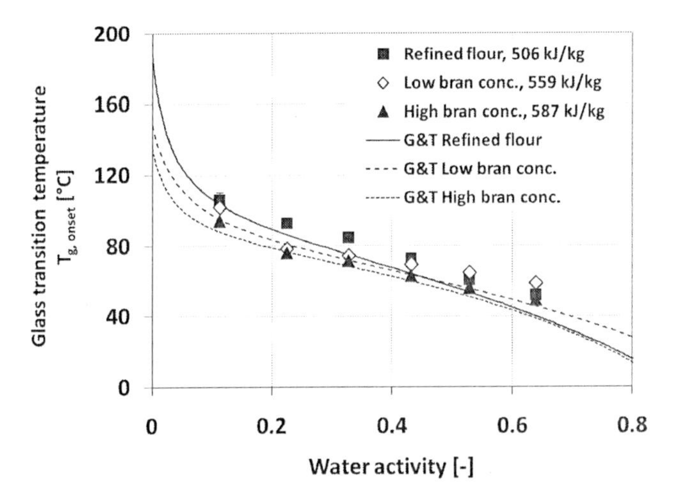

Condition 8 (180 °C, 22 % and 800 rpm)

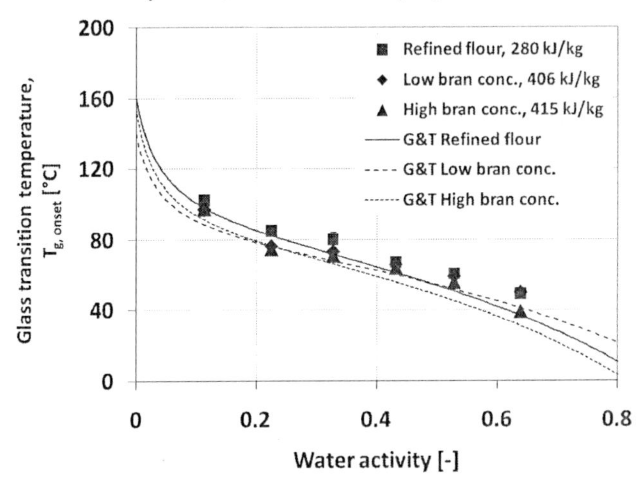

Fig. 4.15: **Glass transition temperature $T_{g,\,onset}$ depending on the water activity a_w of samples with increasing bran concentration extruded at condition and at condition 8 (specific mechanical energy (SME) is indicated next to the sample name)**

IV.2.6. Effect of bran on starch transformations during extrusion

The extent of starch transformation according to the process conditions and bran content may be explained using the solid state diagrams of the unprocessed material displayed in Fig. 4.16.

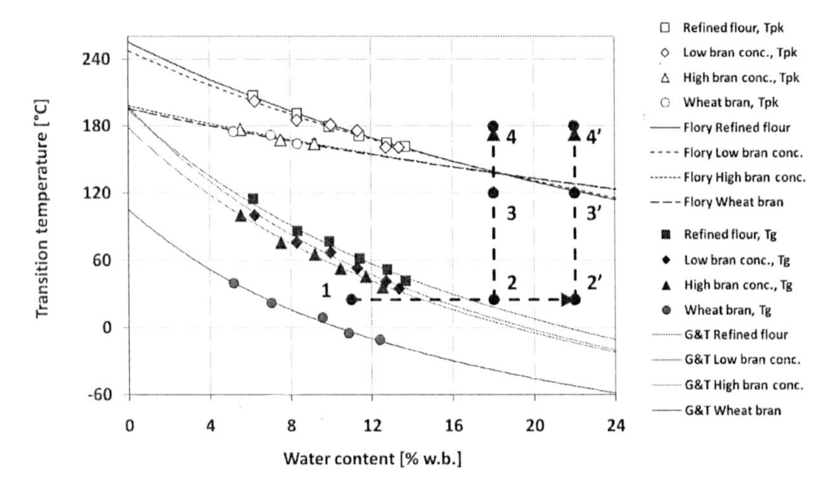

Fig. 4.16: Illustration of the hydrothermal path of extruded samples during extrusion (step 1: raw flour, steps 2 and 2': hydration to 18 % or 22 % water at 25 °C, steps 3 and 3': heating to 120 °C at 18 % or 22 % water, steps 4 and 4': heating to 180 °C at 18 % or 22 % of water)

On this state diagram the variation in melting T_{pk} and glass transition $T_{g,\ onset}$ temperature depending on the water content are reported. In cooking extrusion, a combination of mechanical and thermal energy input transforms semi-crystalline starch into a plasticized, "molten" state, enabling the material to be formed, expanded and digested. Melting the material includes exceeding both the glass transition and melting temperature, with water acting mainly as "plasticizer". The hydrothermal pathway of the samples in the extrusion conditions can be represented in the state diagram in Fig. 4.16. The stage one from conditions 1 to 2 represents the hydration phase of the flour once in contact with fresh water in the extruder, increasing its amount from about 11-13 % of water to 18 % or 22 %. This hydration step is usually represented as an isothermal process (e.g. Kokini et al., 1994). Nevertheless, it is likely that hydration is not instantaneous once water is in contact with the flour. Further migration of water may occur along the extruder barrel length while flour is undergoing physicochemical changes and temperature is increasing. The hydration of the flour allows reaching the glass transition of the flour going from a glassy state to a rubbery stage decreasing the viscosity of the amorphous starch. The second stage from condition 2 or 2' (18 % or 22 % of water content in the feed at room temperature) to condition 3/3' or 4/4' is the increase in temperature of the flour-water mix up to 120 °C or 180 °C. This allows melting the crystalline parts of starch.

The melting temperature at the peak and the glass transition temperature at water contents of 18 % and 22 % were extrapolated using the Flory-Huggins (equation 3.17) and Gordon & Taylor equations (equation 3.18), respectively. The values are shown in Tab. 4.1. At a barrel temperature of 180 °C, the melting temperature is exceeded for all bran concentrations and process conditions (Fig. 4.16 and Tab. 4.1). At the lower barrel temperature of 120 °C, the melt temperature is in the same range as the melting temperature T_{pk} (Tab. 4.1). Nevertheless, no crystallinity remained in all samples (see § IV.2.3). This indicated a melt of the starch crystallites. Only the effect of water on phase transitions is reported on Fig. 4.16, the time-scale effects as well as the energy input would be necessary to have a complete picture of parameters driving the transformations of starch in the extruder.

Table 4.1: Product temperature T_d measured at the die compared to the melting T_{pk} and glass transition temperatures $T_{g,\ onset}$ obtained from the modeling curves (RF: refined flour, LB: low bran concentration, HB: high bran concentration)

	Water content in the extruder [%]	Product temperature (T_d) [°C]	Estimated T_{pk} [°C]	Estimated $T_{g,\ onset}$ [°C]
Refined Flour	18	130 to 171	139	18
	22	118 to 171	121	-2
Low Bran	18	131 to 180	139	8
	22	125 to 169	123	-12
High Bran	18	136 to 176	138	5
	22	127 to 171	128	-14

The loss of starch crystallinity is likely also induced by mechanical stress in the extruder. Amorphous structures able to hydrate, swell and burst under shear in hot water were also found after extrusion (see Chapter IV.2.3). As the extrusion melt temperature was always well above $T_{g,\ onset}$ (Fig. 4.16 and Table 4.1), the molten starch matrix could move and deform freely in the extruder under shear. The presence of such structures may be explained by the high pressure during extruding leading to compaction of the melt at the die exit and preventing their total disruption. The remaining swelling capacity of these structures might be linked to the amylopectin structure after extrusion. Indeed amylopectin was shown to be mainly involved in the swelling capacity of native starch granules (Tester and Morrison, 1990).

With increasing the bran concentration, the matrix glass transition temperature decreases but a difference in starch solubility was found only for the higher bran concentration (see

Chapter IV.2.2). This might indicate a bran concentration threshold from which starch transformation starts to be modulated. This threshold may be attributed to several factors, being additive or counteracting each others:

(i) The reduction in glass transition temperature with bran content will result in a decreasing starch viscosity (Williams et al. 1955). This might therefore reduce mechanical starch transformation in the extruder due to reduced mechanical stress.

(ii) The presence of insoluble bran particles may increase the matrix melt viscosity in the extruder as shown in Chapter V (see Fig. 5.9b). This seems to be confirmed by the higher SME values obtained when increasing the bran concentration (see in Appendices in Tab. 10.5). Local mechanical stress between bran particles and starch molecules may also increase the starch solubility. A decrease in bran particle size would be expected to increase the surface of bran in contact with starch and increase the extent of shear transmitted to the starch molecules. Nevertheless, no significant effect of the bran particle size on the water solubility and absorption indices was observed. This may be explained by the difference in average particle size of the fine and coarse bran which is not large enough to induce significant differences in starch transformation.

(iii) The obtained results were based on the assumption that no further physicochemical change occur after expansion, during cooling or storage. Starch retrogradation can be excluded as no remaining starch crystallinity remained in the samples.

IV.3. Effect of extrusion conditions on bran physicochemical properties

IV.3.1. Content of total, soluble and insoluble dietary fiber

As previously reported, starch properties are significantly modified during extrusion. Nevertheless, depending on the conditions and bran concentration, bran may also be modified during extrusion.

Tab. 4.2: Total (TDF), insoluble (IDF), soluble (SDF) dietary fibers content of unprocessed ingredients and extruded samples at low specific mechanical energy (SME, condition 7, 180 °C, water content in the feed of 18 % and 400 rpm) and at high specific mechanical energy (SME, condition 2, 120 °C, water content in the feed of 18 % and 800 rpm)

	Refined flour			Low bran concentration			High bran concentration		
Fiber content [% d.m]	Not processed	233 [kJ kg^{-1}]	607 [kJ kg^{-1}]	Not processed	271 [kJ kg^{-1}]	630 [kJ kg^{-1}]	High bran conc.	292 [kJ kg^{-1}]	706 [kJ kg^{-1}]
TDF	2.8 ± 0.2	2.8 ± 0.1	2.6 ± 0.5	12.6 ± 0.2	12.4 ± 0.9	13.9 ± 0.5	24.4 ± 0.2	23.5 ± 0.8	25.8 ± 1.7
SDF	0.70 ± 0.04	0.9 ± 0.2	1.2 ± 0.5	1.9 ± 0.1	1.5 ± 0.3	2.2 ± 0.1	2.4 ± 0.5	2.1 ± 0.4	2.3 ± 0.5
IDF	2.1 ± 0.1	1.88 ± 0.0	1.33 ± 0.4	10.8 ± 0.3	10.9 ± 1.2	11.7 ± 0.4	22.0 ± 0. 3	21.4 ± 1.2	22.9 ± 1.4

Table 4.2 summarizes the total dietary fiber as well as the soluble and insoluble dietary fiber content of the unprocessed material and extruded samples at the two extreme conditions of specific mechanical energy input levels. Irrespective of the bran concentration and process conditions, no significant change in total, soluble and insoluble dietary fibers could be observed after extrusion (Tab. 4.2). Gualberto et al. (1997) observed no change in insoluble dietary fiber content, but an increase in soluble fibers after extrusion of wheat bran. For wheat bran, Ralet et al. (1990) reported a significant increase in soluble dietary fibers and a decrease in insoluble dietary fiber content.

Considering the results reported for wheat bran and the ones found in this study, it appears that starch had a protective effect on bran fibers. Nevertheless, Björck et al. (1984) reported a significant increase in soluble dietary fiber and a decrease in insoluble dietary fiber content after extrusion of refined or whole wheat flours, which contains a smaller amount of wheat fibers. However, in all these studies the wheat source and analytical method differed.

IV.3.2. Morphology of bran particles

The extrusion conditions also affected the bran particle size. Fig. 4.17 shows light microscopy images of extruded samples at different levels of bran concentrating and specific mechanical energy input. As shown in samples without added bran (Fig. 4.17a, b) protein particles, stained in green, could be identified and showed a spherical or rectangular shape depending on the process conditions. In the extruded samples containing wheat bran (Fig. 1c, d, e, f), these green protein particles could still be observed. Nevertheless larger particles with varying size and shape depending on the process condition were also stained in green. Their number increased with bran concentration.

Refined flour, condition 7, SME = 233 kJ kg^{-1}

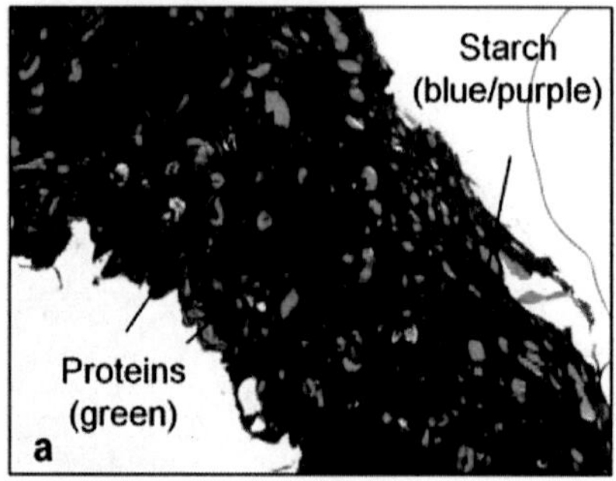

Refined flour, condition 2, SME = 607 kJ kg^{-1}

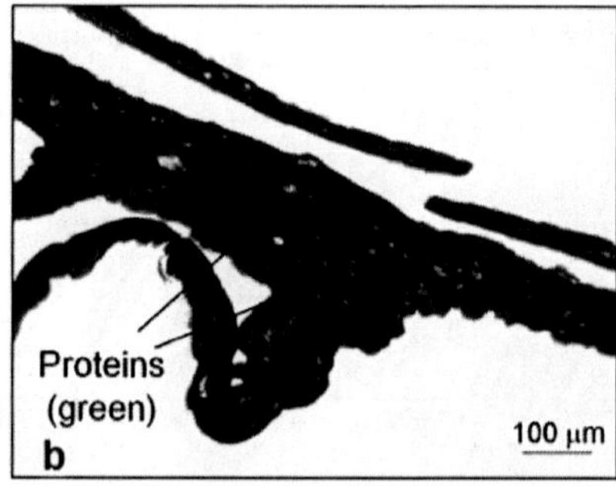

Low bran, condition 7, SME = 271 kJ kg^{-1}

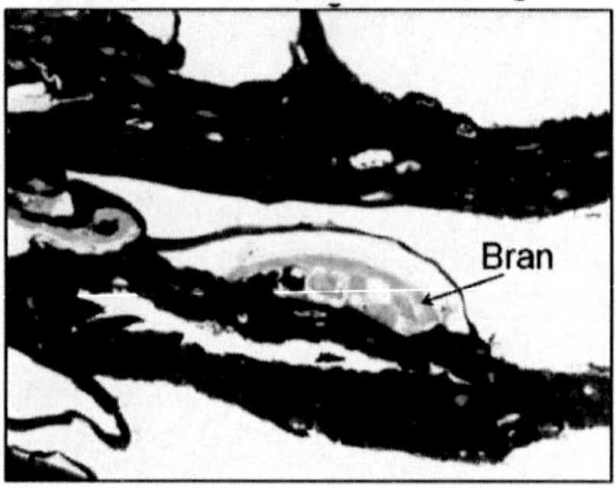

High bran, condition 7, SME = 292 kJ kg^{-1}

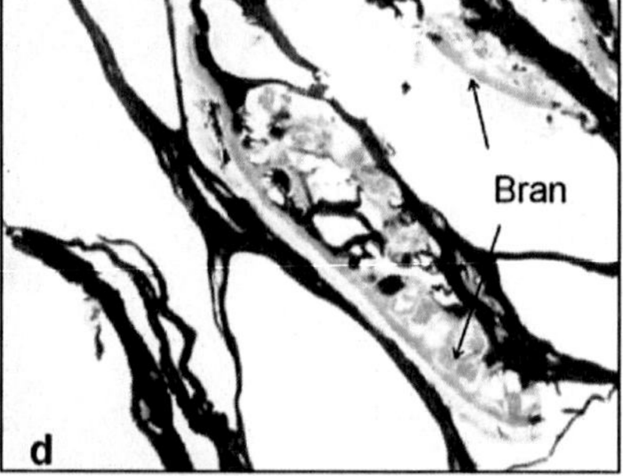

Low bran, condition 2, SME = 630 kJ kg^{-1}

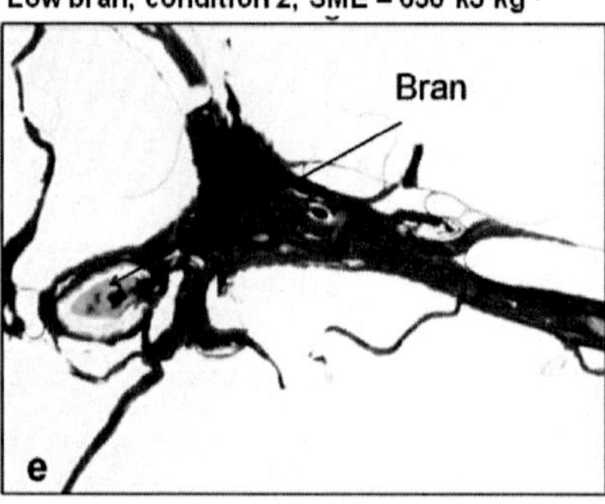

High bran, condition 2, SME = 706 kJ kg^{-1}

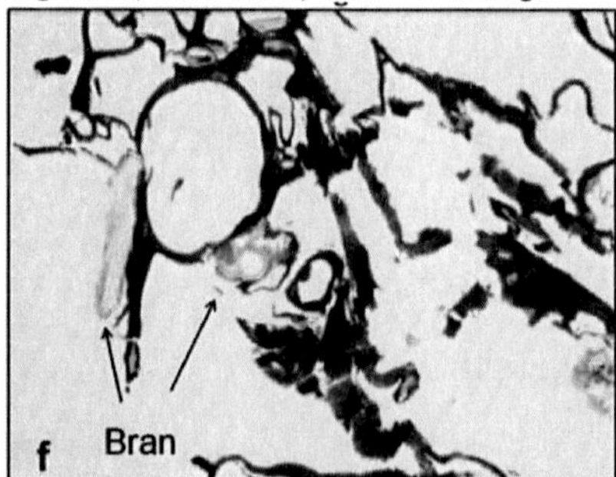

Fig. 4.17: Microscopy images of bran-containing extruded samples at low specific mechanical energy input (SME, condition 7, 180 °C, water content in the feed of 18 % and 400 rpm) and at high specific mechanical energy input (SME, condition 2, 120 °C, water content in the feed of 18 % and 800 rpm)

Due to their large particle size and similarities in shapes and structure with wheat bran particles, showed in published articles using other microscopy techniques such as fluorescence microscopy (Antoine, et al., 2003), they can be identified as bran particles. The staining in green of the bran particles indicates the presence of proteins in their cell walls. At low specific mechanical energy (SME, condition 7 at 180 °C, 22 % feed water content and 400 rpm) almost intact bran particle, mainly constituted of the aleurone layer (proteins contained in bran are stained in green), were observed for the bran containing samples (low bran, Fig. 4.17c and high bran, Fig. 4.17d). At high specific mechanical energy (SME, condition 2 at 120 °C, 18 % feed water content and 800 rpm), smaller particles, corresponding to broken bran particles were observed for the samples containing bran (low bran, Fig. 4.17e and high bran, Fig. 4.17f). The increase in mechanical stress in the extruder may be responsible for this decrease in bran particle size.

IV.4. Conclusions

Starch properties are significantly modified by varying extrusion conditions. Irrespective of the process conditions and bran concentration, the crystalline structure of starch is lost after extrusion. Still, some amorphous structures that could hydrate and burst under shear during rapid visco analysis measurement remain after extrusion. The amount of these structures appears to depend on the process conditions (decreasing with increasing the specific mechanical energy) and the macromolecular state of amylopectin. Increasing the specific mechanical energy decreases the molecular weight of amylopectin, while amylose appears to remain unaffected.

Wheat bran contains mainly dietary fibers. The total, soluble and insoluble dietary fiber content is not significantly modified during extrusion. As shown in this work, the bran particles morphology is modified by extrusion leading to a reduction in particle size. This reduction in particle size is linked with an increase mechanical stress in the extruder. Addition of wheat bran to wheat flour leads to an increase in the water solubility of starch at the highest bran concentration. This likely means an increase in the degree of starch transformation. Nevertheless, no change in molecular size distribution of starch can be detected when increasing the bran concentration.

The higher starch solubility measured in the experiments when increasing bran may come from the increasing mechanical shear between the bran particles and starch and/or the increase in the composite matrix viscosity in the extruder (see Chapter V). Increasing wheat bran concentration decreases the hygrocapacity of the composite matrix. This means that the wheat fibers are more hydrophobic than starch. At same water content, addition of bran decreases the glass transition of starch of nearly 13 K at the highest concentration. This is due to more available free water in the starch phase. The decrease in glass transition of starch may induce a decrease in the viscosity of the starch phase. This may counteract the mechanical stress induced by extrusion or between starch and bran particles. The increase in solubility of starch with the bran concentration may influence the melt rheological properties (see Chapter V) and therefore its expansion properties.

CHAPTER V. MELT RHEOLOGY

V.1. Introduction and hypothesis

In the previous Chapter, it was shown that starch undergoes physicochemical changes during extrusion. It was also reported that starch may be transformed to a higher extent depending on the bran concentration. Starch transformations influence the rheological properties of the melt. The rheological properties of the extruded refined flour should thus also be modified by the addition of wheat bran. To measure the shear viscosity of the melt with increasing bran concentration in the conditions of extrusion, an on-line twin-slit rheometer was built. This rheometer was based on the design of the "Rheopac" system of Vergnes et al. (1993) (see Chapter II.2.4.3.1). Compared to rheometers with a single slit, twin-slit rheometers enable to change the shear rate in the slits while maintaining similar ingredient properties in the extruder. This is of importance as the physicochemical state of starch has an effect on the melt rheology (see Chapter II.2.4.3.1). If a similar thermomechanical history of the ingredients can be achieved in the extruder when using the rheometer or the die, the shear viscosity data, obtained with the rheometer can be compared to the expansion properties when extruding with the die. In this work, the "Rheopac" system was improved by the use of an innovative sliding system to control the slit high (Horvat et al., 2009). This enables to extend the range of shear rates and extruder pressures that can be reached compared to rheometers with a fix slit height. In addition, the temperature in the slits was controlled using separate thermal circulators to limit temperature variation induced by the difference in shear rates between the two.

The extensional viscosity of the melt influences expansion. It may be modified by the addition of bran. The extensional viscosity is obtained by analyzing the flow in a converging die (see Chapter II.4.3.2). In this work the Bagley plot describing the pressure drop at the die entrance depending on the length of the die was established (see Chapter II.4.3.2). Through these experiments, the effect of the bran on the extensional properties of the melt can be estimated.

In this Chapter, the challenges related to the use of the twin-slit rheometer applied to extruded melts with increasing bran concentration are presented and discussed. This includes the maintenance of a similar degree of transformation of the ingredients in the extruder with the rheometer or with the die. The effects of wheat bran on the shear viscosity and Bagley plot of extruded wheat flour are also reported.

V.2. Effect of bran and process conditions on shear viscosity of the melt

V.2.1. Challenges related to on-line measurement of shear viscosity

V.2.1.1. Temperature and pressure instability in and between the slits

Preliminary trials with the twin-slit rheometer were performed by extruding the samples with an increasing bran content at 180 °C with water content in the feed of 22 % and at 400 rpm (condition 7). The shear rate in the slits was varied and the pressure and temperature changes in both slits recorded. Controlling these parameters is important to ensure a reliable measurement of the melt shear viscosity.

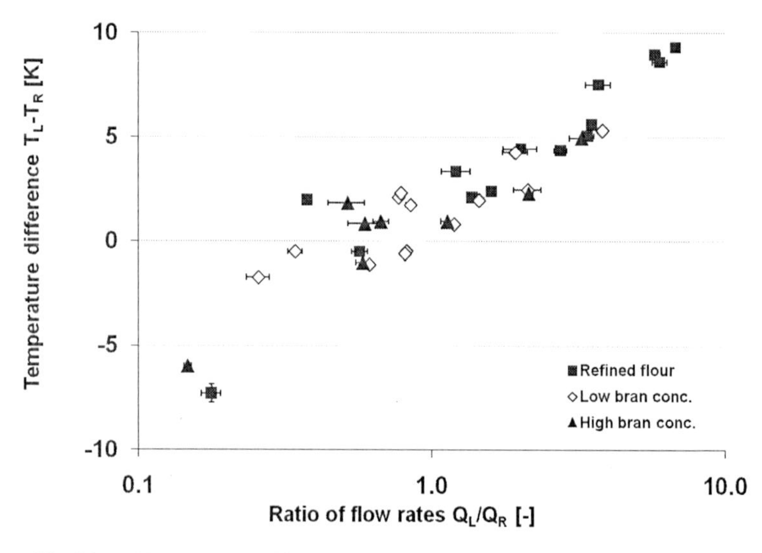

Fig. 5.1: Temperature difference between the left (T_L) and right (T_R) slit as a function of the ratio of flow rates between the left (Q_L) and right (Q_R) slit of samples with increasing bran concentration extruded at condition 7 (180 °C, 22 % water content in the feed and 400 rpm)

The experiments showed that a temperature gradient within and between the two slits was observed and could not be completely eliminated despite the independent cooling circuits.

At constant shear rate, the temperature in the slits was not homogeneous. No clear trend between the slit length and the temperature could be reported. Temperature differences across the slit and the shear rate were also not clearly related. The differences in temperature along the slits were below 5 K (representing less than 4 % of the absolute values of temperatures). The slit temperatures remained affected by the flow rate (Fig. 5.1). The temperatures increased with the flow rate but the differences between the lowest and the highest measured temperatures never exceeded 10 K (representing less than 6 % of the absolute values of temperature) (Fig. 5.1). As shown in Fig. 5.1, at equal shear rates ($Q_L = Q_R$) a higher melt temperature was measured in the left slit (up to 3 K).

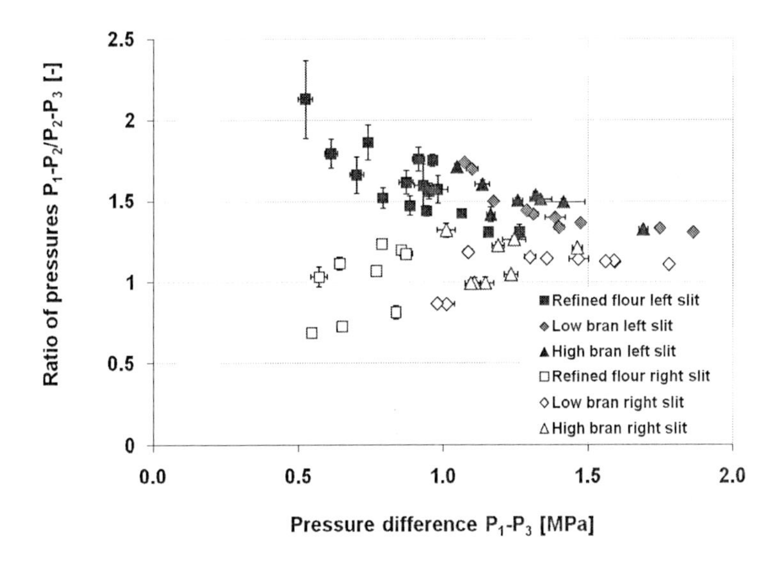

Fig. 5.2: Ratio of pressure differences between P_1-P_2 and P_2-P_3 as a function of pressure drop along the full slit length P_1-P_3 for samples with increasing bran concentration extruded at condition 7 (180 °C, 22 % water content in the feed and 400 rpm)

If the viscosity of the material and thus the shear stress transmitted at the slit wall are independent of the position along the slit, the pressure should decrease linearly along the slit length. The pressure difference measured between the first and second pressure transducer should therefore be identical to the one between the second and third transducer. However, a non linear pressure profile was determined along the slit length. The difference between P_1-P_2 and P_2-P_3 (pressure sensory place along the slit length, see Chapter III.6.1) was negatively correlated to the overall pressure difference in the slit (see in Fig. 5.2). This difference was higher for low P_1-P_3 values (up to 40 % of the full pressure drop between

P_1 and P_3), i.e. lower shear stresses and therefore lower shear rates. This non-linear pressure drop in the slits, especially at low shear rates, may be explained by several factors (which might partially counteract each other):

(i) Material heterogeneity with presence of undisrupted granules that gradually undergo physicochemical changes along the slit,

(ii) Further gradual macronutrients degradation with concomitant changes in viscosity,

(iii) Viscous heating and/or temperature imbalances in the slit leading to a change in viscosity over the slit length,

(iv) Bubble formation (especially at low P_1-P_3 differences) leading to an increase or decrease in viscosity (depending on the gas volume fraction),

(v) Pressure dependence of the melt viscosity.

For similar pressure differences between P_1 and P_3, the left slit was more sensitive to the pressure nonlinearity in the slit than the right slight which showed only a small nonlinearity and for some conditions, at low P_1-P_3 differences, even an opposite effect. At similar shear rates, a higher pressure imbalance between the pressure sensors P_1-P_2 and P_2-P_3 was observed for the left slit. This, as well as the systematically higher temperature previously mentioned for the left slit and pressure imbalance within this slit may also be linked to the design of the rheometer. Since the extruder screw geometries and kinematics (anticlockwise rotation) do not exhibit symmetry with respect to the two slits, one flow may actually be favored. This may result in differences between the left and right slits, as we observed. The higher pressure nonlinearity between the pressure transducers in the left slit, especially at low shear rates (low P_1-P_3 values) also led to significant differences in shear stress values at similar apparent shear rates, as the shear stress was calculated according to the pressure drop between P_1 and P_3. Considering this difference in pressure drop between P_1-P_2 and P_2-P_3, the shear stress will be calculated hereafter using in equation (3.13) (Chapter III) the pressure difference ΔP between the pressure sensor P_1 and P_3 separated by a distance of 90 mm ($2\ x_r = 90$ mm).

Individual temperature control of each slit enabled to reduce the difference in temperature and pressure between the slits. Nevertheless, temperature gradients within and between the slits and associated pressure non linearity along the slits could still be observed. Design of a more efficient heat transfer system between the slit and the cooling system may enable optimizing the measurements.

V.2.1.2. Wall slip assessment

Concentrated fluid suspensions and polymer melts may exhibit slip at solid surfaces. This may result in a lower measured pressure along the slit and underestimation of the shear viscosity. In case of slip at the wall, correction procedures need to be applied as described in Chapter II.4.3.1.

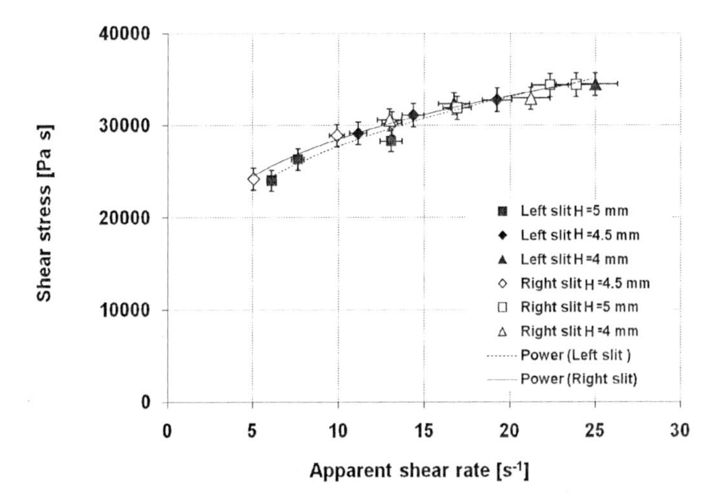

Fig. 5.3: Shear stress vs. apparent shear rate of low bran concentration sample extruded at condition 7 (180 °C, 22 % water content in the feed and 400 rpm), using different slit heights (H) for the left and right slits

In order to investigate slippage at the slit walls, the rheometer slits were operated at three different heights (H = 5, 4.5 and 4 mm) and the shear stress and apparent shear rate were evaluated for each slit height (Fig. 5.3). A systematic difference between values derived for different slit heights would indicate slip at the wall (Makosko, 1994). Such a difference was not detected, irrespective of the bran concentration. Therefore it can be concluded that for the measured shear rates and shear stresses, no slippage occurred.

V.2.1.3. Ingredient thermomechanical history

To ensure that the ingredient thermomechanical history in the extruder is not changed by using the rheometer instead of the die, experiments were performed with the die (index *d*) and then with the rheometer (index *r*). Melt pressure (*P*) and temperature (*T*) as well as specific mechanical energy input (SME) were determined for both set up. When these values are comparable, a similar transformation of the ingredients can be assumed.

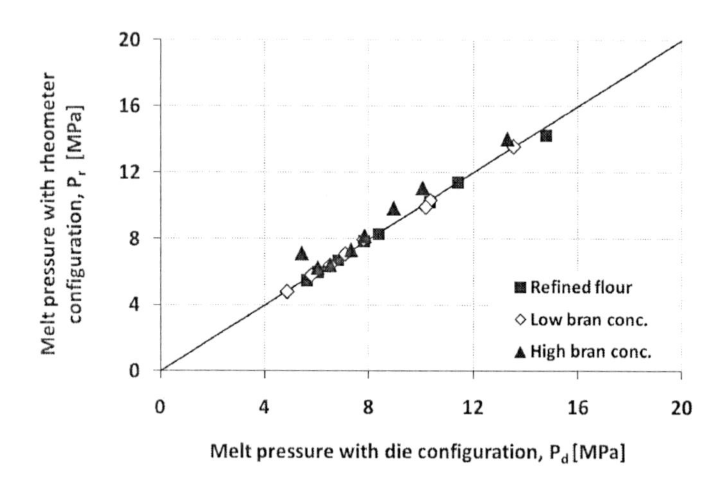

Fig. 5.4: **Comparison of melt pressure measured in the extruder according to different extrusion conditions when equipped with the rheometer (P_r) or with the die (P_d)**

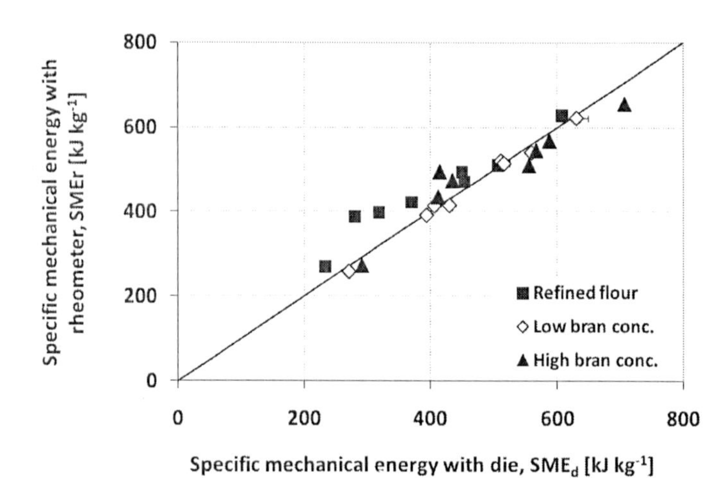

Fig. 5.5: **Comparison of specific mechanical energy measured in the extruder according to different extrusion conditions when equipped with the rheometer (SME_r) or with the die (SME_d)**

The melt pressures with the attached die P_d or with the attached rheometer P_r are compared in Fig. 5.4. Similar back pressures were achieved for both the refined flour and the low bran concentration samples. Nevertheless, significantly higher back pressures were reported for some of the experiments at high bran (HB) concentration when using the rheometer. For these experiments, the melt pressure in the extruder when using the rheometer with slits piston fully open was higher than the melt pressure obtained when using the die. In order to get a suitable range of shear rates, the pistons had to be slightly closed resulting in a higher melt pressure compared to the die configuration.

The specific mechanical energy (SME) values obtained with the die, SME_d, or with the rheometer, SME_r, are compared in Fig. 5.5. A significant difference between SME_d and SME_r was observed for the refined flour (RF). This difference was higher at low specific mechanical energies when the barrel temperature was set at 180 °C. Slight but significant differences between SME_d and SME_r could also be observed for the high bran (HB) concentration.

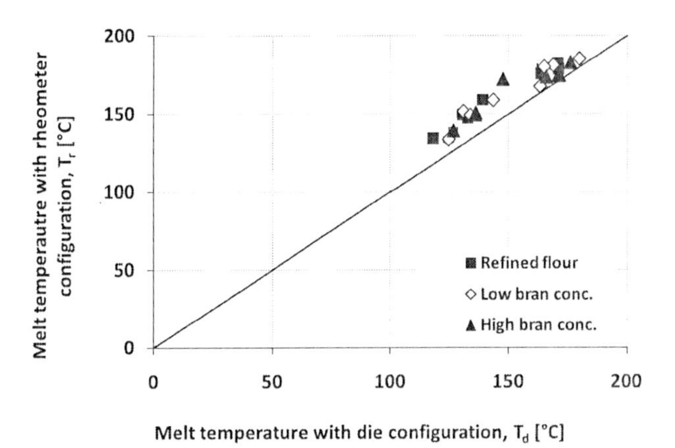

Fig. 5.6: Comparison of melt temperature measured in the extruder according different extrusion conditions when equipped with the rheometer T_r or with the die T_d

The melt temperature with the attached die T_d or with the attached rheometer T_r are displayed in Fig. 5.6. At same process conditions, the melt temperature achieved with the rheometer T_r was significantly higher than the one obtained with the die T_d. The differences in melt temperature when extruding with the die or the rheometer appeared to be lower at high temperatures. Unlike the die, the rheometer was temperature-controlled. The metal temperature (controlled by hot water circulation) was set higher than the product temperature to reach the target temperature T_d in the slits. Energy dissipation coming from the rheometer and transmitted by conduction to the front plate may contribute to the higher melt temperature of the rheometer configuration. The difference in temperature between the rheometer and the die (T_r - T_d) showed a positive trend with the specific mechanical energy (correlation coefficient, r^2 = 0.67, 0.61 and 0.49 for RF, LB and HB, respectively). This indicates that the differences in temperature were driven by the mechanical energy input in the extruder. The higher melt temperatures measured with the rheometer may also be explained by the difference in design between the die (one channel) and the rheometer entrance (separation to form two channels). This may induce higher local stresses and/or residence time in the front plate.

The use of the rheometer may result in further transformation of the ingredients due to the increased residence time and shear stress. This may lead to significant differences in the ingredient physicochemical properties when using the rheometer or the die at similar extrusion conditions. In order to evaluate this, the water solubility index (WSI) was used as an indicator of the ingredient transformations (Guy, 2001). It was measured on the samples extruded with the die and then with the rheometer at same extrusion conditions. As shown in Fig. 5.7, the water solubility index was modified when using the rheometer compared to the die. The degree of change was dependent on the specific mechanical energy input and on the shear rate in the slits. A low specific mechanical energy input (condition 7 at 180 °C, 22 % feed water content and 400 rpm) leads to a significant increase in water solubility index of the refined flour samples at increasing shear rate (Fig. 5.7). The range of shear rates obtained with the bran containing samples was comparable to the one obtained with the refined flour. Nevertheless only slight significant changes could be observed for the bran-containing samples. The observed significant increase in water solubility index was more pronounced at higher specific mechanical energy (as e.g. in condition 2 at 120 °C, 18 % feed water content and 800 rpm). The water solubility index of the samples increases with shear rate, irrespective of their bran concentration (Fig. 5.7). As the shear rates in the slits were significantly different at low and high specific mechanical energy (see Fig. 5.7), it is difficult to explain the more pronounced water solubility changes at high specific mechanical energy. The higher melt shear viscosity measured at condition 2 means a higher mechanical stress (see Chapter V.2.2). This higher mechanical stress may contribute to this difference in water solubility changes.

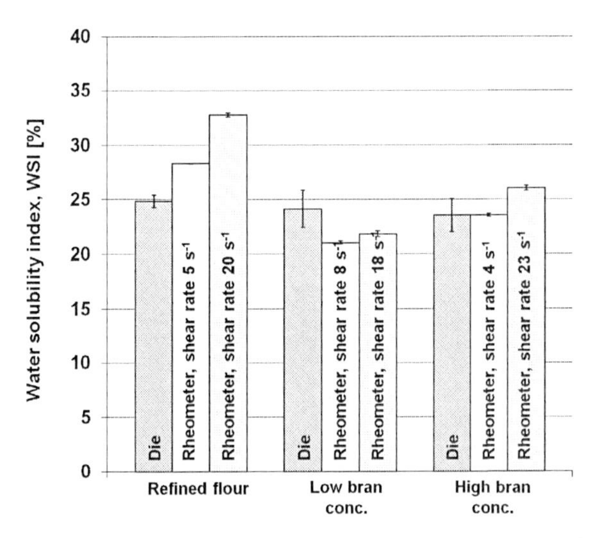

Condition 2 (120 °C, 18 %, 800 rpm) - SME range 600 – 700 kJ kg-1

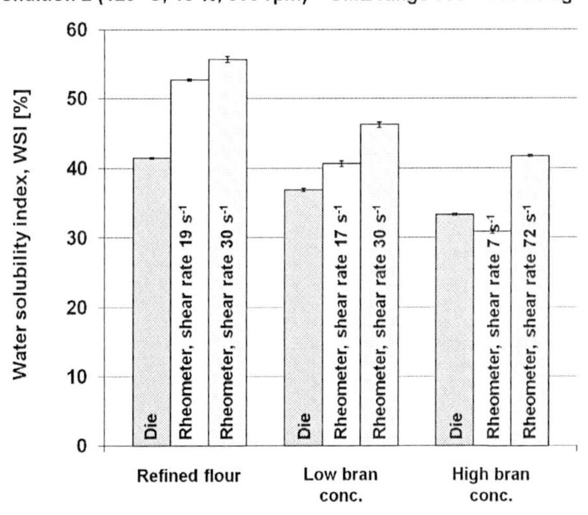

Fig. 5.7: **Water solubility index (WSI) of extruded samples with increasing bran concentration when extruding with the die or with the rheometer at different the shear rates applied in the slits**

The increase in water solubility index occurring in the rheometer may be attributed to starch degradation. Amylopectin branches may be further degradated due to its larger hydrodynamic volume compared to amylose (e.g. Schuchmann and Danner, 2000, Brümmer et al., 2002). As previously reported by Robin et al. (2010a), the refined flour samples appeared to be more sensitive to increasing residence time and shear rate in the rheometer compared to the samples with added bran. This might be due to the higher starch concentration in these samples. Starch is more sensitive to heat and mechanical energy input compared to bran fibers.

Concluding, similar ingredient transformations were obtained when extruding with the rheometer or the die, although some minor modifications may occur in the rheometer. This will allow comparing the rheology data generated with the rheometer to the expansion data obtained with the die. Comparison between the rheology and the expansion data will be discussed in Chapter VI.

V.2.2. Shear viscosity of the melt

V.2.2.1. Flow behavior of the melt

Although some minor technical problems remain, it can be assumed that the adjustable twin-slit rheometer enables a reliable melt shear viscosity measurement. It was applied for the measurement of the melt viscosity with increasing bran concentration and different process conditions.

The viscosity curves of the samples with increasing bran concentration extruded at different process conditions are presented in Fig. 5.8. A range of apparent and corrected shear rates from 5 s^{-1} to 30 s^{-1} and 10 s^{-1} to 100 s^{-1}, respectively, was achieved with the on-line rheometer (Fig. 5.8). The extruded samples exhibited a pseudoplastic behavior as generally reported for extruded cereals (e.g. Harper and others 1971 or Vergnes and other 1987) (Fig. 5.8). The power law index values n for the refined flour ranged from 0.08 to 0.27 and the consistency factor K from 8500 Pa s^n to 46000 Pa s^n (Appendices Tab. 10.5) depending on the extrusion conditions. These power law index values were close to those found by Wang and others (1990) (n = 0.15) for wheat flour using a single slit rheometer. They were nevertheless lower than those reported by Singh and Smith (1999) ($0.21 < n < 0.51$) for wheat flour using a single slit rheometer or by Lach (2006) (n = 0.40) using a twin-slit rheometer. These differences in power law index values may be explained by the difference in starch transformation, and especially of amylopectin degradation as suggested by Della Valle et al. (1996) or Xie et al. (2009).

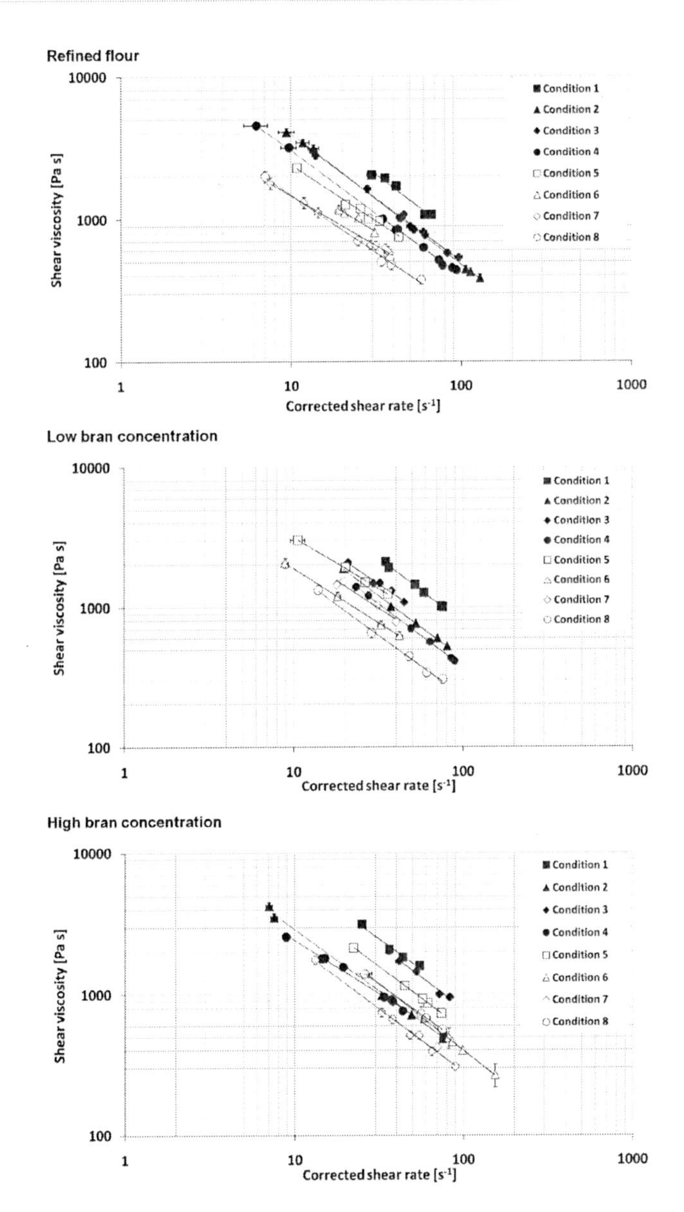

Fig. 5.8: **Viscosity curves of samples with increasing bran concentration at different conditions (see Chapter III, Tab. 3.2 for the extrusion conditions)**

V.2.2.2. Effect of process conditions on shear viscosity of the melt

From the viscosity curves displayed in Fig. 5.8, the effect of process conditions and bran concentration on consistency factor *K*, shear viscosity $\eta(30\ s^{-1})$ and power law index *n* were analyzed. The shear viscosity was calculated according to equation (3.15) (Chapter III) at a corrected shear rate $\dot{\gamma}$ of 30 s^{-1}. This value of shear rate was chosen within the range of corrected shear that could be experimentally measured (no extrapolation to values not experimentally measured was made).

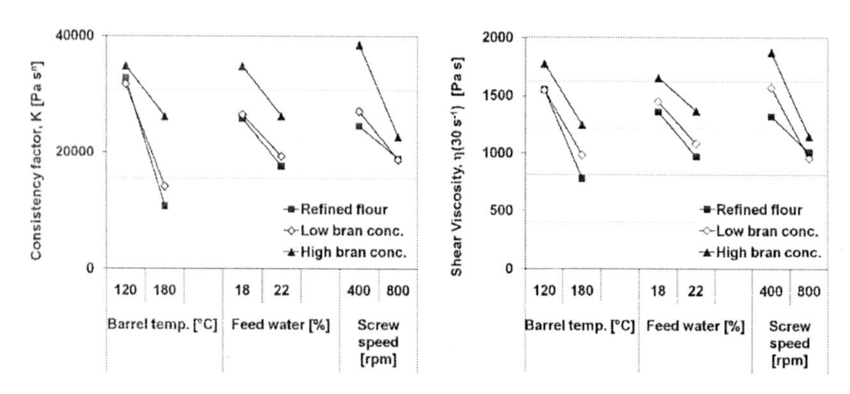

Fig. 5.9: Effect of the process parameters on consistency factor *K* and the shear viscosity (30 s^{-1}) of extruded samples with increasing bran concentration (the distance between two grid lines indicates the least significant difference)

The effect of the process conditions on the consistency factor *K* and power law index, *n* are displayed in Fig. 5.7. The least significant difference (LSD), indicating the minimum difference in values from which the effect is statistically significant was for the consistency factor 15400 Pa s^n, for the shear viscosity 400 Pa s and for the power law index 0.12.

The change in consistency factor value *K* and shear viscosity $\eta(30\ s^{-1})$ with the process conditions showed a similar trend for all the recipes, irrespective of the bran concentration (Fig. 5.9). The correlation factor (r^2) between the statistical model of the experimental plans and the experimental data was high for the consistency factor (r^2 = 0.92 and 0.73) and for the shear viscosity (r^2 = 0.92 and 0.90) at no or low bran concentration (see Appendices in Tab. 10.1). This means that the effect of increasing the level of one process parameter on the shear viscosity or the consistency factor was independent of the level of the two others. At high bran concentration, the correlation factor for the consistency factor and the shear viscosity was lower (r^2 = 0.50 and 0.63, respectively). This indicates that interactions between the process parameters were high and that the effect of one parameter on the shear viscosity and consistency factor was dependant on the levels of the others. This also shows that using an experimental plan was the appropriate approach.

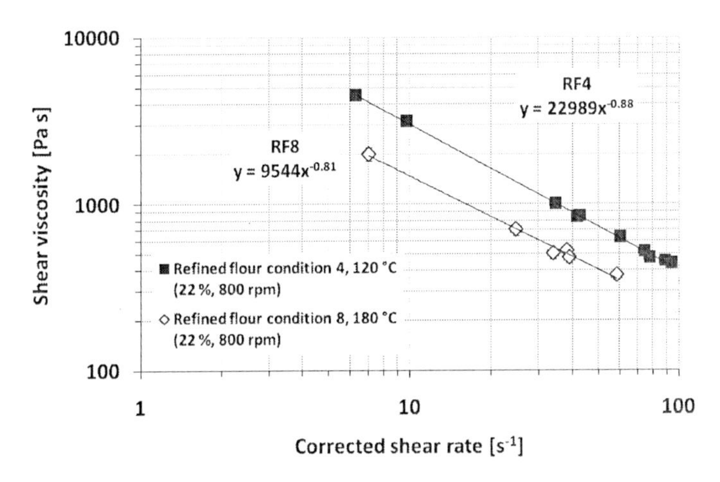

Fig. 5.10: Effect of barrel temperature from 120 °C (RF4) to 180 °C (RF8) on shear viscosity of extruded refined flour at 22 % water content and 800 rpm

The effects of the process parameters showed similar trends on the consistency factor K and shear viscosity $\eta(30\ s^{-1})$ (Fig. 5.9). The barrel temperature had the most significant effect and decreases the shear viscosity (Fig. 5.9). Fig. 5.10 depicts examples of viscosity curves of the refined wheat flour extruded at water content of 22 % and screw speed of 800 rpm, while increasing the barrel temperature from 120 °C (RF4) to 180 °C (RF8). As described through the Arrhenius relationship, viscosity is decreased by increasing the product temperature (e.g. Senouci and Smith 1988).

Increasing the water content in the feed decreased the shear viscosity (Fig. 5.9). Fig. 5.11 gives again exemplarily details showing the viscosity curves of the refined flour extruded at a barrel temperature of 180 °C and a screw speed of 400 rpm, while increasing the water content in the feed from 18 % (RF5) to 22 % (RF7). This is in agreement with the expected plasticizing effect of water on the starch matrix.

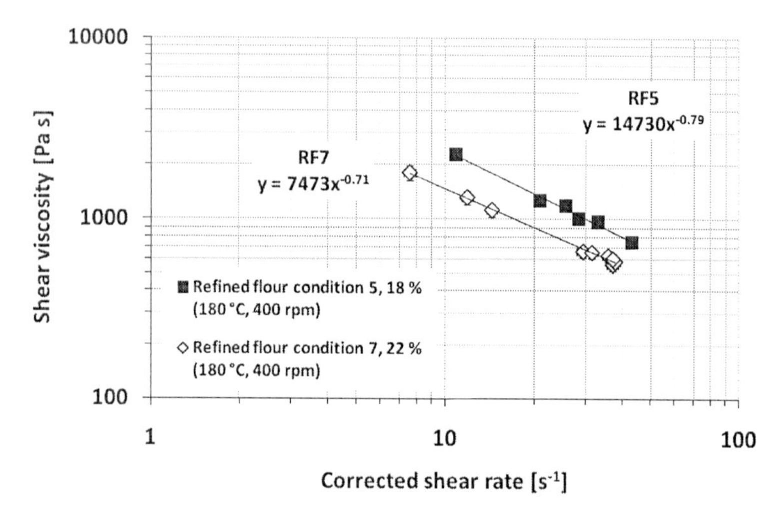

Fig. 5.11: Effect of increasing melt water content from 18 % (RF5) to 22 % (RF7) on shear viscosity of extruded refined flour at barrel temperature of 180 °C and 400 rpm

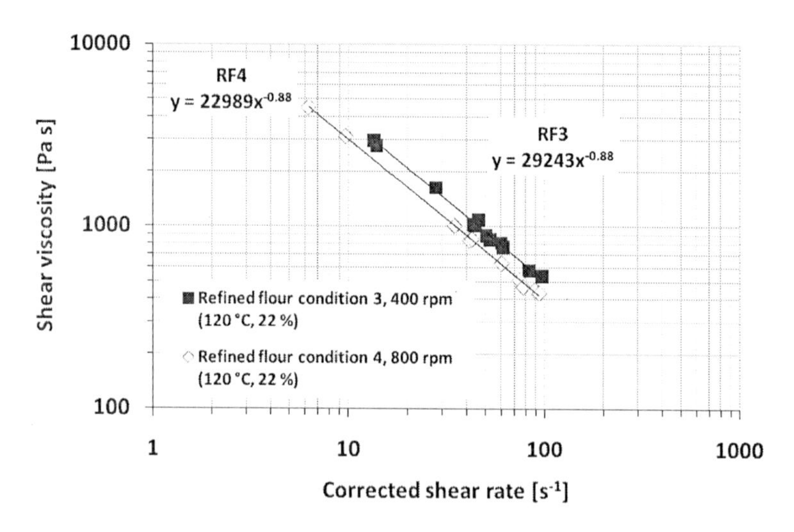

Fig. 5.12: Effect of increasing screw speed from 400 rpm (RF3) to 800 rpm (RF4) on the shear viscosity curves of extruded refined flour at a barrel temperature of 120 °C and water content of 22 %

The screw speed also decreased the shear viscosity values (Fig. 5.9). This can be seen in Fig. 5.12 for extruded refined flour samples at a barrel temperature of 120 °C and water content in the feed of 22 %, while increasing the screw speed from 400 rpm (RF3) to 800 rpm (RF4). The decreasing shear viscosity with screw speed can be explained by the thermomechanical history the ingredients when extruded. As shown in Chapter IV in Fig. 4.3, the water solubility index (WSI) of the recipes was increased with screw speed. The increase in water solubility index in extruded starchy foams is associated with the depolymerization of amylopectin branches (Guy 2001). Increasing shear rate in the extruder thus may increase the starch molecular weight which for polymers is correlated to a decrease in the matrix shear viscosity. A similar effect of the ingredient thermomechanical history on shear viscosity of the melt was previously reported e.g. Senouci and Smith (1988a), Parker et al., (1989) or Horvat et al. (2009). The water solubility index (WSI) was decreased with the barrel temperature and the water content in the feed (Fig. 4.3). This decrease in starch transformation may also affect (likely counteract) the decrease in shear viscosity when increasing the barrel temperature and the water content in the feed.

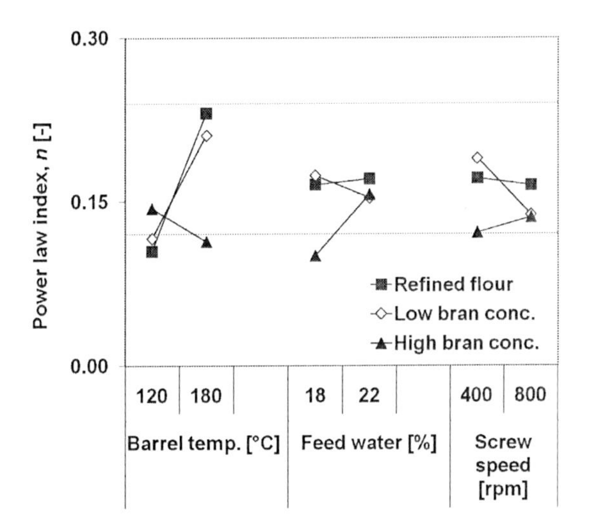

Fig. 5.13: **Effect of the process parameters on melt power law index (*n*) of extruded samples with increasing bran concentration (the distance between two grid lines indicates the least significant difference)**

No clear trend could be observed between the process conditions and the power law index n (Fig 5.13). Only the power law index of the refined wheat flour recipe increased significantly with increasing barrel temperature (Fig 5.13). This means that the pseudoplastic behavior is decreased and the material has a less marked pseudo plastic behavior at higher temperature. A similar effect was previously reported for waxy maize starch (Della Valle and others 1996; Willett and others 1997) and for maize starch (Xie and others 2009; Willett and others 1995).

V.2.2.3. Effect of bran concentration on shear viscosity of the melt

Viscosity curves of the melt extruded at same process conditions but with different bran concentration are shown in Fig. 5.14.

The effect of bran on the consistency factor and shear viscosity of extruded wheat flour was dependant on the process conditions. For instance, at a barrel temperature of 180 °C, a water content in the feed of 18 % and a screw speed of 400 rpm (condition 5), the shear viscosity of the refined flour melt (RF5) was significantly lower than the bran-containing melts (LB5 and HB5 samples) (Fig. 5.14). Whereas, at higher water content in the feed (22 %) and screw speed (800 rpm) (condition 8) the increase in shear viscosity was only significant at the high bran concentration (HB8) (Fig. 5.14). On the other hand, there was no significant difference in shear viscosity at a lower barrel temperature (120 °C), water content in the feed of 22 % and screw speed of 800 rpm (condition 4) when increasing the bran concentration (Fig. 5.14).

The effect of bran on the shear viscosity $\eta(30\ s^{-1})$ and the consistency factor of the refined flour recipe, depending on the process parameters, showed similar a trend and is summarized in the statistical analysis included in Fig. 5.9. The shear viscosity was similar for the refined flour and low bran recipes (Fig. 5.9). It was significantly higher with increasing the bran to the highest concentration, but only at high barrel temperature and low screw speed (Fig. 5.9). Otherwise, shear viscosity was affected by both, process parameters and bran concentration, making an interpretation of the data difficult. The effect of the bran concentration on the power law index n was not significant and did not show a clear trend (Fig. 5.13). Moore et al., 1990 and Wang et al., 1990 observed no effect of wheat bran at concentrations of 8 to 16 % on the apparent viscosity of cooked dough. However the composition of the bran was not specified. More recently, Pai et al. (2009), using capillary viscometry, reported an increase in the apparent viscosity of extruded corn meal supplemented with corn bran to 26 % of total dietary fiber content. In this study, wheat bran had no significant effect at low concentration on the shear viscosity of wheat flour. This appears consistent with the results of Moore et al. (1990) and Wang et al. (1990). At higher concentration, the effect of bran on the shear viscosity of the melt was significant. Therefore, it is likely that a threshold concentration of bran exists. From this threshold concentration the shear viscosity of the composite matrix starts to be affected.

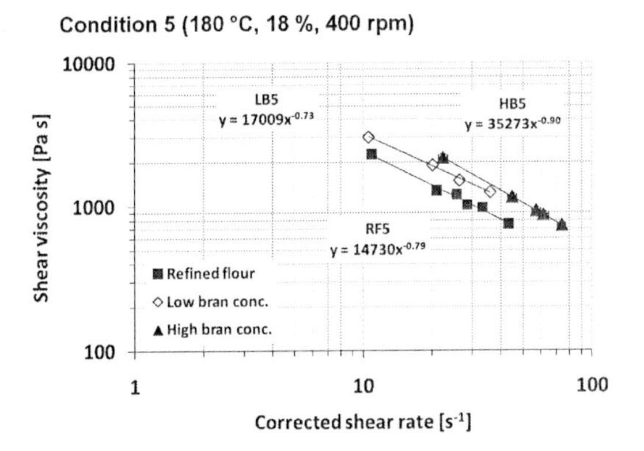

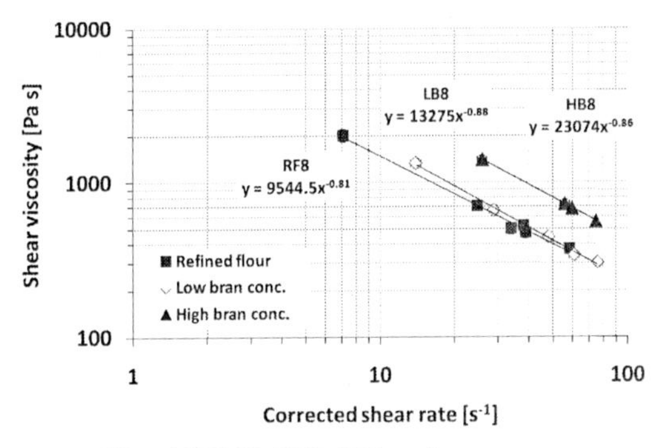

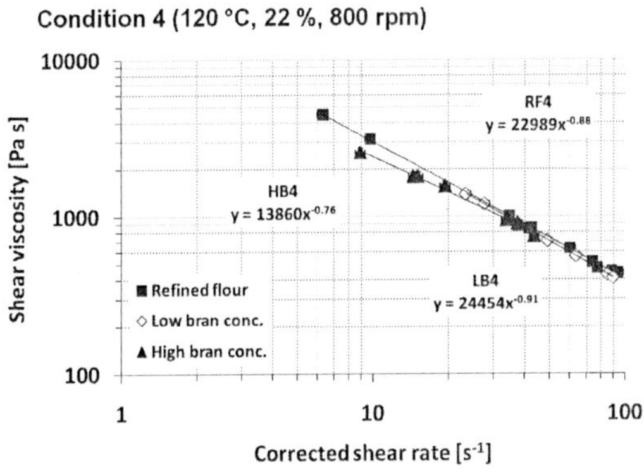

Fig. 5.14: Effect of bran concentration on shear viscosity of the melt extruded at different conditions to extrusion

Wheat bran mainly contains fibers being mostly insoluble. Their solubility was not increased after extrusion in the conditions of this study (see Chapter IV in Tab. 4.2). Bran may be considered as a filler of the continuous starch phase. Indeed definition of fillers in the field of plastic extrusion is given to materials that retain their structures during processing and remain inert, insoluble and thermally stable. They also have no phase transitions, no catalytic activity and low additive adsorption (Hohenberger, 2009). According to this definition wheat bran appears to fit most of these criteria. As reported for synthetic polymers reinforced with fillers, fillers increase the matrix shear viscosity and shear thinning behavior (Xanthos 2005). This was for instance shown for organic or inorganic fillers in extruded polystyrene by Lobe et al. (1979) and Lou and Harinath (2004). The effect on shear viscosity is more pronounced at low shear rates due to the formation of structures networks. Higher shear rates tend to orient fibers tend to orient fibers and flakes to different degrees depending on their size, rigidity, concentration and interactions with the matrix (Xanthos, 2005). Therefore, at high shear rates the characteristics of the continuous matrix tend to drive the shear viscosity of the matrix. At lower shear rates, the properties of the fillers tend to govern the shear viscosity of the composite matrix. Fillers may have a limited effect at low concentration and then more strongly affect the composite rheological properties as the concentration increases (Shenoy, 1999). At high concentration the interactions between fillers strongly influence the matrix rheological properties. These interactions between the particles may be hydrodynamic forces, excluded-volume, frictions between particles or mechanical (Shenoy, 1999). They could also influence the orientation of the fibers during rotation in the extruder, in the die or in the rheometer slits (Shenoy, 1999). Additionally to the volume fraction and interactions between the bran particles, adhesion properties at the boundaries with the starch phase, particle size distribution, aspect ratio (ratio between the largest dimension to the smallest) and orientation of the bran particles may also affect the shear viscosity of the composite matrix. If we consider wheat bran particles as fillers of the continuous starch matrix, their effect at low concentration may not be enough to increase the shear viscosity of the composite matrix. Only at higher bran concentration, their effect on the free volume of the starch phase may lead to an increase in shear viscosity of the composite material.

Tab. 5.1: Glass transition temperature $T_{g,\,onset}$ at extrusion condition 1 (120 °C, water content in the feed of 18 % and 400 rpm) and condition 8 (180 °C, water content in the feed of 22 % and 800 rpm)

	Water content in the extruder [%]	$T_{g,\,onset}$ at condition 1 [°C]	$T_{g,\,onset}$ at condition 8 [°C]
Refined flour	18	15.1	21.3
	22	-4.7	2.8
High bran conc.	18	12.1	5.5
	22	-4.8	-12.4

The glass transition temperature of the extruded samples at conditions 1 (120 °C, 18 % water content in the feed and 400 rpm) and 8 (180 °C, 22 % water content in the feed and 800 rpm) was estimated from the extrapolation of the G&T model (see equation 3.18 in Chapter III) and according to the water content in the extruder (Tab. 5.1). The T-$T_{g,\ onset}$ difference would increase up to 16 K when increasing the bran content to the highest bran concentration at constant feed water content and melt temperature T (Tab. 5.1). The lower glass transition of starch when adding bran may result in a lower starch viscosity at constant melt temperature (Williams et al., 1955). This lower viscosity of the starch phase may counteract the increase in shear viscosity of the matrix due to bran addition.

At higher bran concentration, the effect of bran particles on the shear viscosity was not the same depending on the process conditions (Fig. 5.9). This depicts interacting effects between the process conditions and the bran concentration. The interactions between the bran concentration and the process conditions may be explained by the following phenomenon:

(i) As reported in Chapter IV (Fig. 4.4), the extent of starch transformation (according to the estimated starch water solubility) appeared to be higher when increasing the bran concentration to the highest level. The higher water solubility of starch was depending on the process conditions. This may decrease the shear viscosity of starch and counteract the effect of bran on the matrix shear viscosity.

(ii) In this work, using the "coarse" or "fine" bran quality had no significant effect on the shear viscosity when extruded at same conditions (see Chapter III.2). This may be explained by the small difference in particle size distribution between these two fractions (see Appendices in Fig. 10.1). Nevertheless, according to the microscopy images shown in Chapter IV, Fig. 4.17, the bran particle size after extrusion was reduced at high specific mechanical energy input. The bran particle size also appeared significantly lower than the average particle size of the unprocessed bran. Decreasing the bran particle size during extrusion may reduce the free volume of molten starch and increase its shear viscosity. The effect of this reduction in particle size on the shear viscosity may be higher at low shear rate. Indeed, as earlier mentioned, the properties of the filler tend to govern the shear viscosity of the matrix at low shear rate (Xanthos, 2005).

(iii) The orientation of the bran particles which may be changed by bran-bran interactions and distribution in the melt of the bran particles, which may not be homogeneous, should also be considered. This may influence the melt viscosity.

(iv) Wheat bran is not fully inert. Indeed, it was shown that it exhibits an elastoplastic rheological behavior (Peyron et al., 2002; Antoine et al., 2003; Grefeuille et al., 2007; Hemery et al., 2010). A small glass transition was even detected between -40 °C and -50 °C (Hemery et al, 2010). Additionally, the mechanical properties of bran may be changed according to the process conditions. Hemery et al. (2010)

recently reported that the mechanical properties of wheat bran are significantly affected by temperature and water content. They reported an increase in the extensibility and a decrease in stiffness of bran with water and temperature. The degree in extensibility and stiffness was different through the bran layers depending on the layer thickness, composition and cross-linking (Peyron et al., 2002; Antoine et al., 2003).

V.2.2.4. Mathematical fitting of experimental values

V.2.2.4.1. Mathematical fitting of shear viscosity data depending on process conditions

Mathematical fitting of the experimental data will enable predicting the shear viscosity of the melt depending on the process parameters. Several mathematical fittings of experimental viscosity data have been proposed in the literature for modeling the effect of process parameters and thermomechanical history of the ingredients on the shear viscosity of the melt. These different models are presented in the Chapter II.2.4.3.1.

The mathematical fitting proposed by Vergnes and Villemaire (1987) considers all the process parameters studied in this work and fits best the obtained data (see equation 5.3).

$$K = K_0 \exp\left[\frac{\Delta E}{R}\frac{1}{T} - \alpha MC - \beta SME\right] \qquad (5.3)$$

Here, T is the temperature (in Kelvin) and MC is the normalized moisture content (from 0 = 0 % water content to 1 = 100 % water content). K_0, $\Delta E/RT$, α and β are constant parameters obtained when fitting the equation to the experimental data and using the solver function of Excel.

In equation (5.3), the temperature plays a major role. Measuring it exactly in the experiments is thus crucial. Efforts were therefore made to adjust and maintain the temperature of the melt in the two rheometer slits similar. Nevertheless, as shown in this Chapter V.2.1.1 some differences could still be noticed. Therefore, the average temperature measured between the two pressure sensors recording the pressure drop in the slits was preferred to the melt temperature measured at the die plate. As mentioned by Della Valle et al. (1996), the model based on the Arrhenius law is more valid at temperatures well above the glass transition temperature (T_g) ($T >> T_g + 100$ °C). In this work, the glass transition temperature ranged between -12 °C to 21 °C depending on bran concentration and moisture content (see Tab. 5.2). This is well below the melt temperature (ranging from 118 °C to 176 °C, see in Appendices in Tab. 10.5). The power law index n was not significantly modified by the process conditions (Fig. 5.13). Only for the refined flour (RF) and variations of the barrel temperature the power law index was significantly changed (Fig. 5.13). Therefore, it does not really make sense to calculate the power law index according to a linear model as proposed by Della Valle et al. (1996). Experimental values of the power law index were therefore used to predict shear viscosity η at a shear rate $\dot{\gamma}$ = 30 s^{-1} according to equation (5.4):

$$\eta(\dot{\gamma}) = K\dot{\gamma}^{n-1} \qquad (5.4)$$

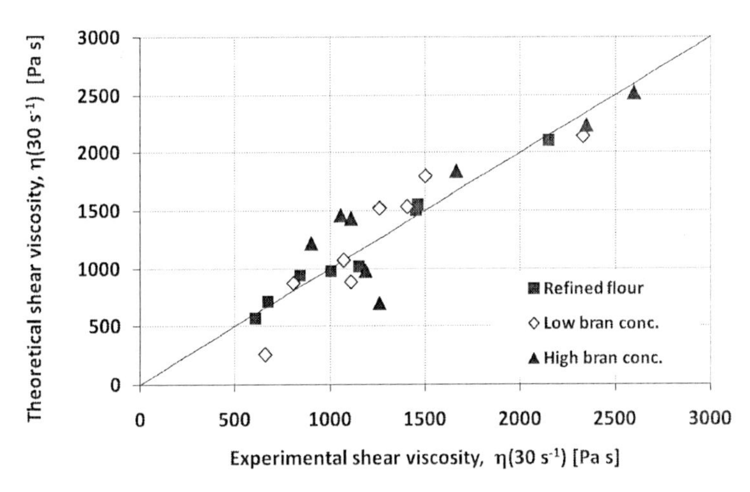

Fig. 5.15: Shear viscosity values obtained from the mathematical fitting (equation 5.4) at $\dot{\gamma}$ = 30 s^{-1} against experimental shear viscosity ($\dot{\gamma}$ = 30 s^{-1}) of extruded samples with increasing bran concentration

Plotting of predicted viscosity against the experimental values is obtained and shown in Fig. 5.15. A satisfactory correlation between the shear viscosity, assessed experimentally, and the one given by the mathematical fit was found for the refined flour (r^2 = 0.92).

Tab. 5.2: Mathematical fitting parameters of shear viscosity data

	K_0 [Pa s]	$\Delta E/R$ [K]	α [-]	β [t kWh^{-1}]	n experimental [-]	r^2
Refined flour	0.081	7343	23.5	0.006	0.08-0.27	0.92
Low bran	0.160	7553	27.4	0.007	0.09-0.27	0.84
High bran	0.002	11201	34.5	0.027	0.08-0.24	0.74

The fitting parameters of the refined flour shown in Tab. 5.2 were in the same order of magnitude as those reported by Della Valle et al. (1996) for maize starch containing 23.3 % amylose (K_0 = 0.03 Pa s, E/R = 7190 K, α = 19.1 and β = 0.0033 t kWh^{-1}). Similar amylose content is found in wheat flour indicating the influence on the shear viscosity of the melt of the amylose fraction in cooked cereals.

The correlation between the experimental and predicted shear viscosity values was also generally satisfactory for the low bran concentration samples (r^2 = 0.84). The correlation was nevertheless poorer at high bran concentration (r^2 = 0.74) (Fig. 5.15). The changes in shear viscosity of the products containing no or only a low concentration of bran can be explained by the individual process parameters. The interactions between the process parameters however, do not allow for a good mathematical fit at high bran concentration.

V.2.2.4.2. Mathematical fitting of shear viscosity depending on bran concentration

The mathematical fitting proposed by Krieger and Dougherty (1959), shown in equation (5.5), was applied to consider the effect of the bran volume fraction on the shear viscosity of the composite matrix at constant process parameters and melt shear rate.

$$\frac{\eta_s\left(30\,\mathrm{s}^{-1}\right)}{\eta\left(30\,\mathrm{s}^{-1}\right)} = \left(1 - \frac{\phi}{\phi_{\max}}\right)^{-\alpha} \qquad (5.5)$$

$\eta_s(30\mathrm{s}^{-1})$ the shear viscosity of the composite matrix containing wheat bran and $\eta(30\mathrm{s}^{-1})$ the one of the refined flour (RF). ϕ is the dry volumetric fraction of the bran particles calculated from the percentage of fibers and considering than bran particles are mostly composed of fibers. ϕ_{max} is the maximum packing volume of fraction. The theoretical value of 0.74 for face centered cubic packing was used. α was 2.5 as suggested by Roscoe (1953) for concentrated solutions. The plotting of the predicted values of shear viscosity depending on the bran concentration and the experimental values is shown in Fig. 5.16.

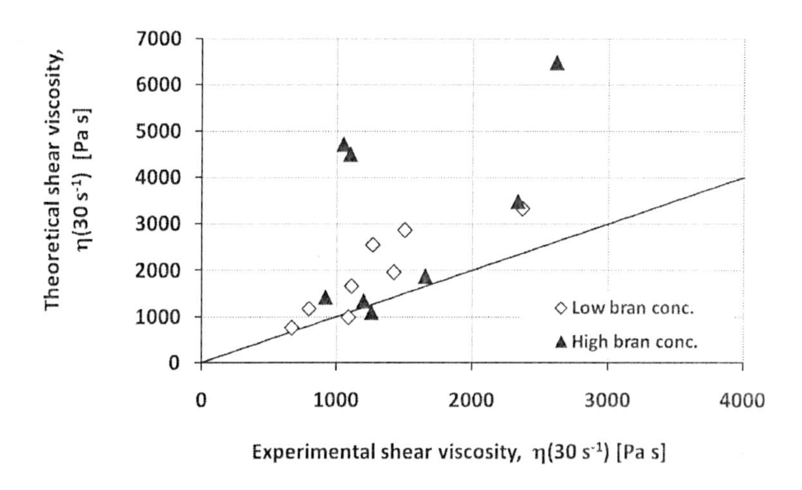

Fig. 5.16: **Shear viscosity values obtained from the mathematical fitting (equation 5.5) at $\dot{\gamma}$ = 30 s^{-1} against experimental shear viscosity ($\dot{\gamma}$ = 30 s^{-1}) of extruded samples with increasing bran concentration**

The correlation factor (r^2) between the experimental values of shear viscosity and the predicted ones for the low bran concentration was satisfactory with a correlation factor r^2 = 0.78. The correlation was much poorer for the high bran concentration samples with a correlation factor r^2 of 0.27. Discrepancies between the predicted and experimental values were higher at high viscosity. These results suggest that other parameters than the bran particle size volume fraction such as bran particle morphology changes (e.g. size, aspect ratio, number) according to the extrusion conditions or interactions between the bran particles and starch may also influence the shear viscosity of the composite matrix, especially at high bran concentration.

V.3. Effect of bran on entrance pressure drop

The samples with increasing bran concentration were extruded at similar extrusion conditions (150 °C, water content in the feed of 20 % and 600 rpm). These conditions were chosen at intermediate values of screw speed, water content and barrel temperature used in the experimental plan in Chapter III in Tab. 3.2. The single slit rheometer described in Chapter III.6.2. was used. The length of the die attached at the end of the rheometer was varied. The pressure drop in the single slit rheometer was measured.

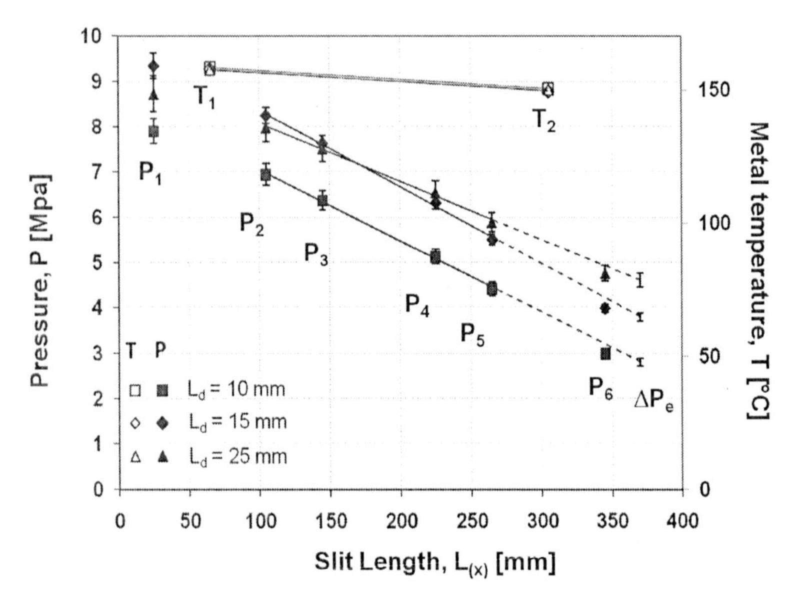

Fig. 5.17: Pessure (P) and temperature (T) over the rheometer slit length ($L_{(x)}$) with different die lengths (L_d). Data are shown for extruded refined flour at 150 °C, water content in the feed of 20 % and 600 rpm

Fig. 5.17 shows the pressure profile P plotted against the length $L_{(x)}$ of the slit with increasing the die L_d/D_d ratio. The shear rate in the slit was 48 s^{-1}. As it can be seen in this figure, ΔP increased when increasing the length of the die. This indicates increased resistance against flow in the slit rheometer due to increased surface of contact/shear in the cylinder die. As reported for the twin-slit rheometer (see Chapter V.2.1.1), the pressure was not linearly decreasing in the slit. The same explanations given in Chapter V, § V.2.1.1 for the twin-slit rheometer can be applied to the pressure drop instability in the single slit, i.e.

(i) Material heterogeneity with presence of undisrupted granules that gradually undergo physicochemical changes along the slit,

(ii) Further gradual macronutrients degradation with concomitant changes in viscosity,

(iii) Bubble formation (especially at low pressure in the slit) leading to an increase or decrease in viscosity (depending on the gas volume fraction),

(iv) Pressure dependence of the melt viscosity and/or

(v) Viscous heating and/or temperature imbalances in the slit leading to a change in viscosity over the slit length.

Temperature imbalance in the slit was observed in the slit as shown in Fig. 5.17. The temperature of the rheometer decreased across the slit length. This may therefore mean a decrease in the melt temperature. While the temperature at the entrance of the slit was generally higher when increasing the bran concentration, it was not significantly different at the end of the slit rheometer. The temperature decrease in the slit was caused by dissipation of heat along the slit. The temperature of the melt coming from the extruder and entering the slit was higher (160-165 °C) than the temperature of the metal part of the rheometer, set to 150 °C. The final temperature at the exit of the slit and entrance of the die was 150 °C, regardless of the bran concentration and die length.

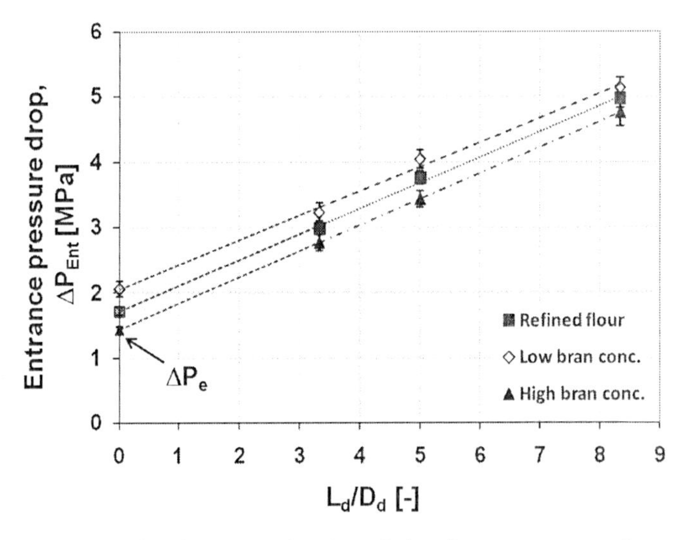

Fig. 5.18: Bagley plot representing the variation of entrance pressure drop at the die ΔP_{ent} depending on the ratio die length L_d over die diameter D_d of samples with increasing bran concentration extruded at 150 °C, water content of 20 % and 600 rpm

The entrance pressure drop at the die was obtained by extrapolating the pressure drop in the slit measured between P_2 and P_5 to a length L of 370 mm corresponding to the entrance of the die (see Fig. 5.17 and Chapter III.6.2). The pressures P_1 and P_6 measured close to the entrance and exit of the slit. They were excluded from the estimation of the entrance pressure drop to avoid influence of flow disturbances at the exit and entrance of the rheometer (Fig. 5.17). This procedure was repeated for the different bran concentrations and die lengths.

The Bagley plot (Bagley, 1957), representing the pressure at the die entrance ΔP_{Ent} as a function of the ratio between die length and diameter L_d/D_d (with D_d having a constant value of 3 mm) of the die is shown in Fig. 5.18. ΔP_{Ent} increases with L_d/D_d. No significant differences in water solubility index (WSI) could be observed depending on the ratio L_d/D_d (see in Tab. 5.3). This indicates that no major changes in the ingredient physicochemical properties occurred depending on the length of the die.

Tab. 5.3: Water solubility index (WSI) of extruded samples (150 °C, water content in the feed of 20 % and 600 rpm) with increasing bran concentration depending on the ratio between die length (L_d) and die diameter (D_d)

	$L_d/D_d = 3.33$	$L_d/D_d = 5.00$	$L_d/D_d = 8.33$
Refined flour	31.8 ± 2.9	31.8 ± 2.1	31.8 ± 1.1
Low bran	29.9 ± 0.2	29.9 ± 0.7	29.9 ± 0.7
High bran	26.4 ± 0.2	26.4 ± 0.2	26.4 ± 0.3

Linear extrapolation of ΔP_{Ent} to a ratio between the die length and its diameter L_d/D_d to 0 (orifice die procedure, see Chapter II.4.3.2) enables to determine the end pressure ΔP_e of a die with virtually no length (see Fig. 5.18). This pressure drop obtained at $L_d/D_d = 0$ can be attributed to the effect of extensional viscosity at the die entrance. The value of ΔP_e was respectively 1.71 ± 0.05 MPa, 2.06 ± 0.12 MPa and 1.43 ± 0.05 MPa for the refined flour, low bran and high bran melts. At low bran concentration this value was significantly increased while it was significantly decreased at the highest bran concentration. The significant increase in ΔP_e when increasing the bran concentration may be explained by the reduced free volume of the starch phase. The lower free volume of starch reduces its capacity to elongate. An increase in end pressure ΔP_e was for instance reported for extruded polystyrene with an increasing concentration of glass fibers (Chan et al., 1978) or polypropylene filled with short glass fibers (Crowson et al., 1980). Unlike for polymers further increasing the bran concentration to the highest level significantly decreased ΔP_e. The decrease in ΔP_e means that less work is necessary for the deformation of the macromolecules corresponding to the flow acceleration in the die. Extruded polymers show higher viscosity and cohesion than starch in conditions of extrusion. They also likely show higher adhesion with synthetic fillers than starch with bran particles. Further increasing the bran concentration is likely reducing the free volume of the starch matrix. Nevertheless, it also decreases the cohesion of the starch matrix, reducing starch-starch interactions and thus leads to ruptures in the starch melt. Due to acceleration of the flow at the die entrance, the molecules are deformed and the starch network may be ruptured. A low cohesion material may require less work for deformation of the macromolecules. This may explain the pressure decrease ΔP_e at high bran concentration. Such phenomena may be similar to synthetic polymer rupture obtained at a maximum shear rate. It corresponds to the rupture of the gross physical structure of the chain agglomerates consequent to shear-induced disentanglement of the elastomer chains or segments thereof, without however involving any chain scission due to the shear action (Gosh et al., 1997). The corresponding shear at which the polymer failure occurs was shown to be reduced when increasing the concentration of fillers such as carbon black in synthetic polymers (Gosh et al., 2000). This was attributed to fewer interactions between the polymer chains leading to disentanglement. This theory might apply to starch when increasing the bran concentration. The increased in starch transformation and lower starch transition temperature at the highest bran content (see Chapters IV.2.5) may also affect the extensional viscosity of starch. Bran-bran interactions, change of orientation at the die entrance, breakage of the bran particles as well as interfacial slippage between the bran particles may also need to be taken into account.

In literature, Pai et al. (2009) reported an increase in the extensional flow of extruded corn flour when increasing corn bran to a content of up to 26 % of dietary fibers. In their study, the extensional flow was measured using lubricated squeeze flow viscometry. Increasing the volume fraction of filler, regardless of shape was reported to significantly reduce the elasticity of the melt. This was reflected by a reduced extrudate swell and concomitant effects on normal stress differences (Xanthos, 2005). A reduction in normal stress differences, corresponding to the reduction in elastic properties of the melt was for instance reported by Lobe and White (1979) for polystyrene melt reinforced with carbon black or by Minagawa and White (1976) for polyethylene melts filled with titanium dioxide. It is likely that bran particles reduce the elastic properties of extruded starch and the die swell. For a detailed explanation more experimental results at different bran concentrations are required.

V.4. Conclusions

With the developed twin-slit rheometer, it is possible to measure the shear viscosity of the melt in the conditions of extrusion. The shear viscosity data obtained from this rheometer can be used to explain the expansion properties obtained with a die and when varying bran concentration and process conditions. The viscosity of the extruded melt shows a pseudoplastic shear thinning behavior, whatever the process conditions and the bran concentration. Under the tested conditions, the shear viscosity of the melt is significantly reduced when increasing the water content, screw speed or barrel temperature. During extrusion, the physicochemical properties of starch are modified depending on the process condition. Changes in the supramolecular and molecular structure of starch also contribute to the changes in the matrix rheological properties.

Increasing the bran concentration increases the shear viscosity only at the highest bran level and at low screw speed, high water content and high barrel temperature. At low concentration the effect of bran is not high enough to induce a significant difference in the shear viscosity of the melt. At higher concentration, the effect on the matrix shear viscosity is more significant and may be further enhanced by bran-bran particle interactions. According to these findings and the ones reported in the literature, it appears that shear viscosity of starchy melts starts to be affected by bran only from a threshold concentration. This is consistent with the existence of a threshold value of filler concentration in extruded synthetic polymers reinforced with fillers. The shear viscosity of the matrix is driven by interactions between the bran concentration and the process conditions. This could be explained by the structural changes of bran particles depending on the process condition. It could be shown that fiber particles are mechanically disrupted at high shear conditions. As by this the specific surface area of bran increased, starch-bran interactions are affected. Mathematical fittings of the experimental shear viscosity data depending on the barrel temperature, water content and specific mechanical energy were established. They enable to predict shear viscosity data within the range of tested conditions. Mathematical fitting of shear viscosity data depending on the volume fraction of bran is difficult to the interactions between starch and bran particles. For this, a more complex mathematical fitting involving these interactions would need to be established.

At same process conditions, the entrance pressure drop of extruded refined flour is significantly modified by the addition of bran. Increasing the bran concentration to the low level significantly increases the die entrance pressure drop. This may mean a higher melt extensional viscosity induced by the reduction in free volume of the melt. At the higher bran concentration the entrance pressure drop is significantly lowered compared to both the refined flour and low bran concentration. Although increasing the bran concentration may further reduce the starch free volume it may also lead to rupture of the starchy melt. This may be due to missing adhesion properties between starch and bran at high bran concentration. Both these effects may lead to a reduction in the melt capacity to elongate. Further work is nevertheless necessary to conclude on the effect of wheat bran particles on the extensional viscosity of wheat flour melts.

CHAPTER VI. EXPANSION PROPERTIES

VI.1. Introduction and hypothesis

The study of expansion has to be divided into its mechanistic steps to understand the effect of the process conditions and bran concentration on expansion and resulting microstructure. The mechanics steps of expansion are nucleation, bubble growth and collapse of gas bubbles inside the melt (see Chapter II.4). All these steps are depending on the rheological properties of the melt. As shown in the previous Chapter, the shear viscosity of the melt was significantly increased only at the highest bran level. The effect of bran on the shear viscosity of the melt was process condition-dependant. The entrance pressure drop of the melt was also significantly modified by the addition of bran, reflecting a change in the extensional viscosity/elasticity of the melt. These changes in the melt rheological properties may explain the different bulk expansion of the samples obtained when increasing the bran concentration.

The rheological properties of the melt govern the growth of the bubbles. The final dimensions, shape and porosity of the samples depends on the bubble (size, shape and density) and cell wall characteristics (thickness) resulting from the expansion mechanism. Therefore, a relationship may exist between the bulk expansion properties and the cellular structure characteristics. Establishing such a relationship will enable to predict the cellular structure characteristics from the bulk expansion properties of the samples. This may allow gaining analytical time and reducing costs as the microstructure is obtained by complex and time-consuming analytical tools obtained with fast and easy techniques.

The effects of bran concentration on the mechanistic steps of the expansion mechanism are reported in this Chapter. The rheology data obtained in Chapter V, at similar extrusion conditions as the expansion properties obtained with the die, are used to explain the effect of bran on each step of the expansion mechanism. The relationship between the bulk expansion properties and the cellular structure is also established and discussed.

VI.2. Effect of bran on bubble nucleation

The gas bubble structure can be directly related to the number of bubbles nucleating at the die (Moraru and Kokini (2003). The nucleus density could be linked to the measured number of cells in the final product assuming that one cell was generated by one nucleus and that bubble coalescence and shrinkage had only a limited effect on the cell density.

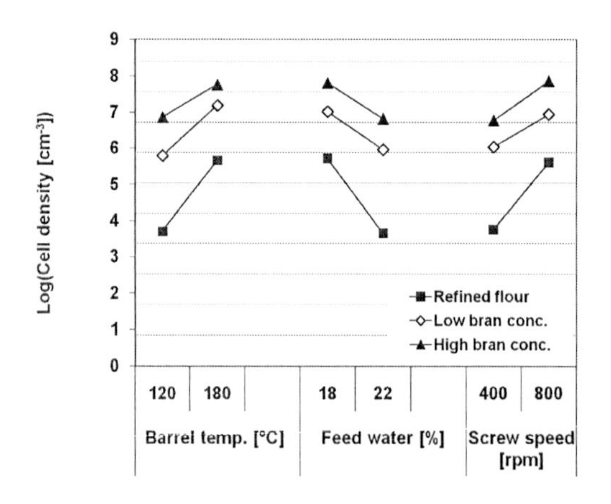

Fig. 6.1: Effects of process conditions on cell density (log scale) of extruded samples with increasing bran concentration

The cell density, depending on the extrusion conditions and bran concentration, was analyzed by X-ray tomography (Fig. 6.1). Regardless of the bran concentration, increasing the barrel temperature and the screw speed increased the number of cells, while a decreased cell density could be observed when increasing the water content. Several theories of nucleation have been discussed for synthetic and natural polymers and may involve heterogeneous, homogenous, cavitation, shear stress-induced nucleation and/or air bubble entrapment in the extruder (see Chapter II.4.2). The exact mechanisms are not yet fully understood. Surely, all these theories may apply in parallel and influence the measured structure, even though some process conditions or ingredients may favor one or the other.

Increasing the bran concentration at similar process conditions increased the cell density N_c following an exponential relationship (see Appendices in Table 10.6). This corresponds to an increase in the degree of nucleation when increasing bran concentration. Bran is composed of fibers that are mostly insoluble. Their solubility was reported in Chapter V (see in Tab. 4.2) and remains unchanged after extrusion. Fibers and starch have a different physicochemical compatibility as shown by the microscopy images in Chapter IV in Fig. 4.17 and by the reduced hygrocapacity of the samples when increasing the bran concentration (see sorption isotherms in Chapter IV in Fig. 4.9). This increased nucleation degree can be associated with the presence of the nonwetted suspended bran particles acting as nucleating particles (Lee, 2000). Physical nucleators which are incompatible with the polymer continuous phase, such as synthetic fibers, are likely to favor heterogeneous nucleation (Lee, 2000). The surface of the bran particles may also be rough or porous. Pre-existing vapor cavities can be located at the surface. Once a critical melt pressure is reached, they may be activated and favor nucleation and bubble growth. Such a phenomenon was previously reported for talc supplementation in extruded plastic foams (Leung, 2009).

Depending on the bran concentration, the change in the process conditions did not show the same quantitative effect on the cell density (thus on nucleation). For instance, raising the temperature from 120 °C to 180°C increases, on average, the cell density from 140 cm^{-3} to 430 cm^{-3} (about 3 fold) for the refined flour (RF). For the high bran concentration the cell density is increased from 1600 cm^{-3} to 2700 cm cm^{-3} (less than 2 fold) (see in Fig. 6.1). Similar observations can be made when changing the other process parameters. Interacting effects between the bran and starch may explain such phenomena. They may involve changes in the bran properties during extrusion. For instance the bran particle size was decreased and its density was increased with the shear stress in the extruder (see Chapter IV, Fig. 4.17). At constant bran volume fraction, the increase in bran particle density, resulting from their breakage during extrusion, may increase the number of sites for nucleation. The increase in bran concentration was also shown to increase the shear viscosity (see Chapter V, Fig. 5.9b). This may increase the shear-stress induced nucleation (see Chapter II.4.2).

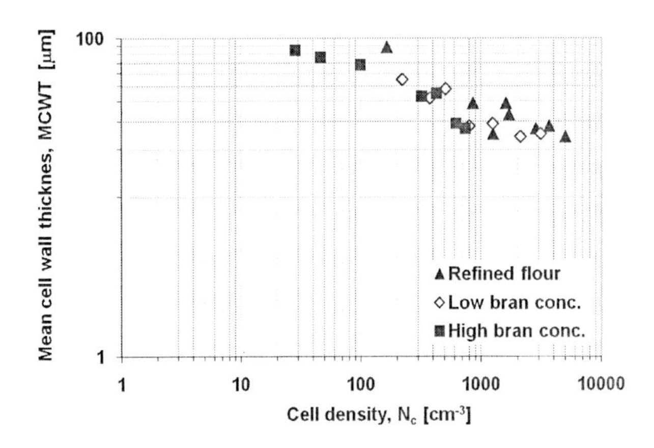

Fig. 6.2: Mean cell wall thickness (MCWT) as a function of cell density (N_c) for samples with increasing bran concentration and extruded at different conditions

At same process conditions, increasing nucleation by increasing the bran concentration may lead to a higher number of growing bubbles in the melt. This may affect the expansion properties. A negative power law relationship was observed between the mean cell wall thickness (MCWT) and the cell density as shown in Fig. 6.2. This implies that the thickness of the bubble membrane was lower for a higher degree of nucleation. This can cause less resistance to the pressure inside the bubble during growth and an increased final bubble dimensions. A higher degree of nucleation might be also cause a higher shear rate at the interface between the growing bubbles. The melt showed a pseudoplastic behavior (see Chapter V, Fig. 5.8). This would reduce its viscosity depending on the conditions. Nucleation also depends on temperature and pressure difference between the melt pressure and saturated vapor pressure at this temperature (see Chapter II.4.2). Therefore, thermal diffusion may be of importance in the nucleation process. It may be modified by the inclusion of bran particle in the continuous starch matrix due to change in the thermal conductivity of the composite matrix.

VI.3. Effect of bran on bubble growth and collapse

VI.3.1. Bulk expansion properties and relationship to cellular structure

VI.3.1.1. Effect of process conditions on bulk expansion and cellular structure

The relationship between the bulk expansion properties (VEI, SEI and LEI) and the cellular structure was investigated. For this an expansion chart, representing the variations of the longitudinal expansion index (LEI) with the square root of the sectional expansion index (SEI) was established.

The expansion chart of the samples with increasing bran concentration is shown in Fig. 6.3. This chart enables to estimate the preferred orientation of the expandate during expansion (sectional, longitudinal or with no preferred direction: isotropic). Pictures of the expanded samples can be seen in Appendices in Fig. 10.2. For cylindrical dies, isotropic expansion would result in a square relationship between the longitudinal and sectional expansion indices ($SEI^{0.5}$ = LEI) (Bouzaza et al., 1996). For refined flour in most of the experiments sectional expansion is dominating ($SEI^{0.5}$ >> LEI). Expansion of the melt was favored towards the *y* and *z* directions and was restricted in the direction of extrusion (*x* direction) (Fig. 6.3).

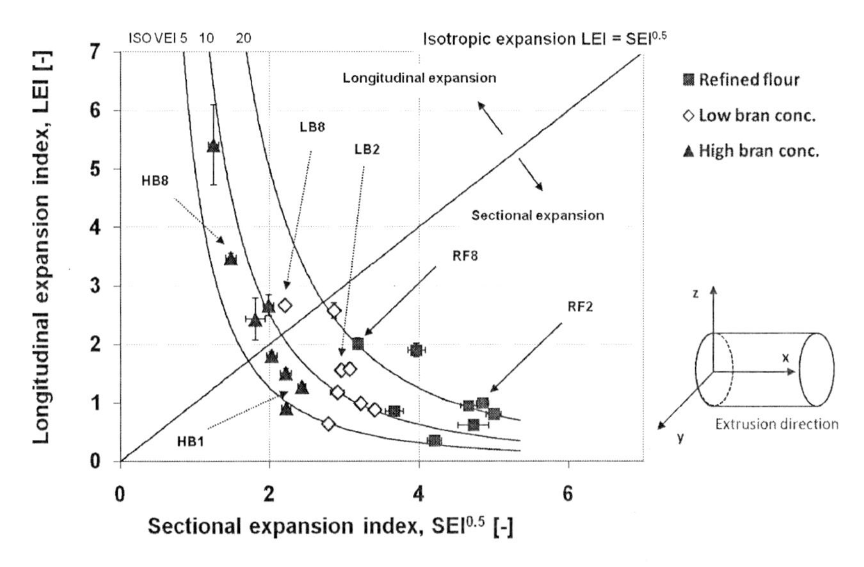

Fig. 6.3: Expansion chart representing the variations of longitudinal expansion index (LEI) with the square root of the sectional expansion index (SEI) of samples with increasing bran concentration

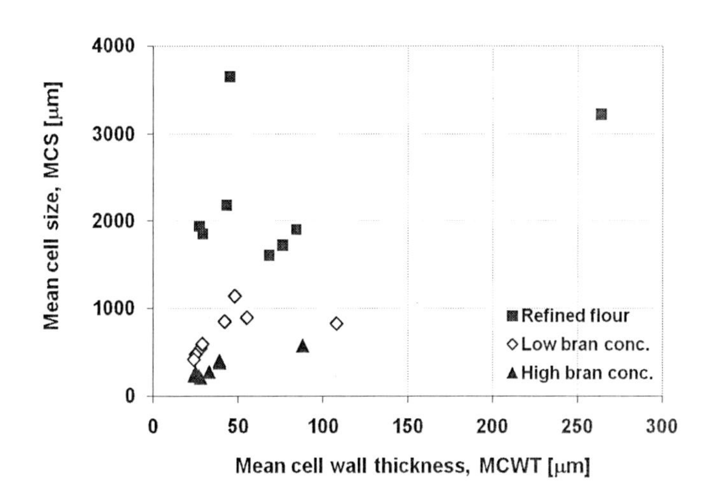

Fig. 6.4: Mean cell size (MCS) vs. mean cell wall thickness (MCWT) of samples with increasing bran concentration extruded at different conditions

The mean cell size (MCS) and mean cell wall thickness (MCWT) of the samples extruded at different conditions and bran concentration were determined by X-ray tomography. Their relationship is showed in Fig. 6.4. Examples of cross sectional and cross longitudinal X-ray tomography images of the extruded samples are shown in Appendices in Fig. 10.4 and 10.5, respectively. A positive trend between the mean cell size (MCS) and the mean cell wall thickness (MCWT) was found for the bran-containing samples (Fig. 6.4). This means that for both low and bran concentration samples, larger cells were associated with thicker cell walls. For the refined flour, two populations of cellular structures with positively correlated mean cell wall thicknesses (MCWT) and mean cell size (MCS) could be observed. Such a positive correlation between MCWT and MCS was previously reported by Babin et al. (2007) for extruded maize starch while no clear relationship between the mean cell size and mean cell wall thickness was observed by Robin et al. (2010a) for extruded wheat flour.

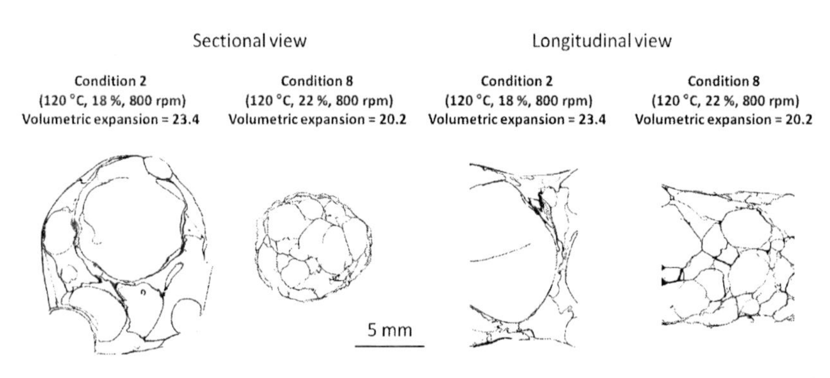

Fig. 6.5: Cross sectional and cross longitudinal X-ray tomography images of refined flour samples having a similar volumetric expansion index

The relationship between the bulk expansion properties and the cell characteristics was assessed comparing samples having similar porosities (and therefore similar volumetric expansion). The refined flour samples extruded at condition 2 (RF2, 120 °C, 18 % water content in the feed and 800 rpm) and at condition 8 (RF8, 180 °C, 22 % water content in the feed and 800 rpm) exhibited close volumetric expansions (VEI = 23.4 vs. 20.2). The sample extruded at condition 8 showed a bulk expansion closer to isotropy with a significant lower sectional (SEI = 10.1 vs. 23.5) and a significant higher longitudinal expansion (LEI = 2.00 vs. 1.00) than the one obtained at condition 2 (RF2) (see in Fig. 6.3). As shown in Fig. 6.5, the sample extruded at condition 8 had a finer structure with a significant higher number of smaller cells (N_c = 740 cm^{-3} vs. 430 cm^{-3}, MCS = 1940 μm vs. 3650 μm) and significant smaller cell wall thicknesses (MCWT = 30 μm vs. 45 μm). It also included more spherical cells than the sample extruded at condition 2 (RF2), having cell shapes closer to oblate ellipsoids (Fig. 6.5).

VI.3.1.2. Effect of bran concentration on bulk expansion and cellular structure

Increasing the bran concentration at similar process conditions significantly decreased the volumetric expansion (VEI). As shown in Fig. 6.3, this was associated with a significant decrease in sectional expansion (SEI) and a stronger acceleration of the melt at the die outlet (increase in LEI). This resulted into bulk expansions closer to isotropy or even favored longitudinal expansion for some high bran samples. The relationship between the volumetric or sectional expansion indices with the bran/fiber concentration followed a linear or a power law relationship (see in Appendices in Tab. 10.6). The longitudinal expansion index (LEI) was increased with the bran/fiber content following a linear or exponential relationship (see in Appendices in Tab. 10.6). An example of these relationships at extrusion condition 4 (180 °C, 18 % water content in the deed and 400 rpm) is shown in Fig. 6.6.

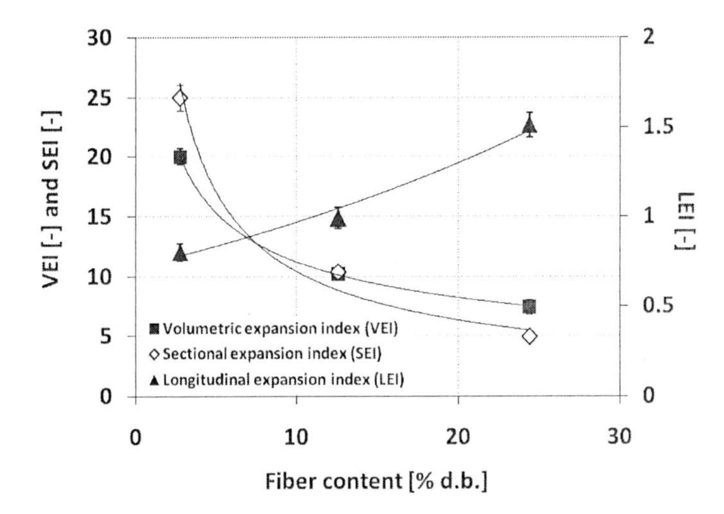

Fig. 6.6: Fiber content vs. expansion properties of samples extruded at condition 4 (180 °C, 18 % water and 400 rpm) (LSD_{VEI} = 2.5, LSD_{SEI} = 3.4, LSD_{LEI} = 0.51)

As shown on the expansion chart in Fig. 6.3 varying process conditions when extruding bran-containing samples was not sufficient to reach the sectional expansions (diameters) obtained with the refined flour.

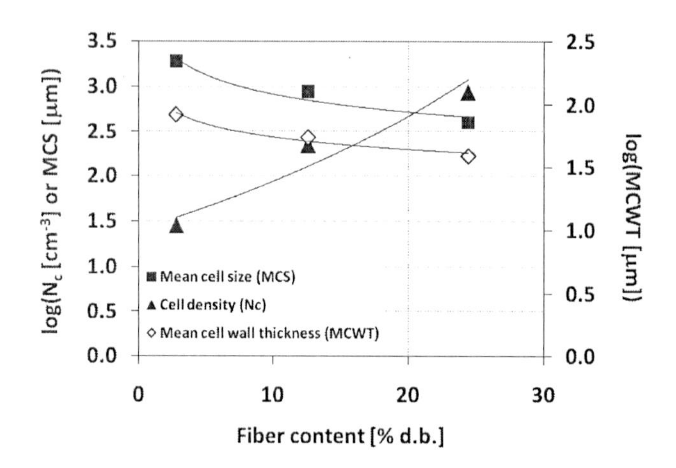

Fig. 6.7: **Mean cell size (MCS), mean cell wall thickness (MCWT) and cell density (N_c) as a function of fiber content of samples extruded at condition 4 ($LSD_{log(MCS)}$ = 0.50; $LSD_{log(MCWT)}$ = 0.20 $LSD_{log(Nc)}$ = 0.84)**

Increasing the bran concentration (hence the fiber content) also significantly decreased the mean cell size (MCS) and the mean cell wall thickness, (MCWT) and significantly increased the cell density (N_c) (see in Appendices in Tab. 10.6). Similar observations were reported by Moore et al., 1990 and Lue et al. (1990) and Yanniotis et al. (2007). The relationships between the bran concentration and the mean cell size or the mean cell wall thickness were well fitted with a negative power law or logarithmic relationship. An exponential function better described the link with the cell density. An example of these relationships at extrusion condition 4 is given in Fig. 6.7.

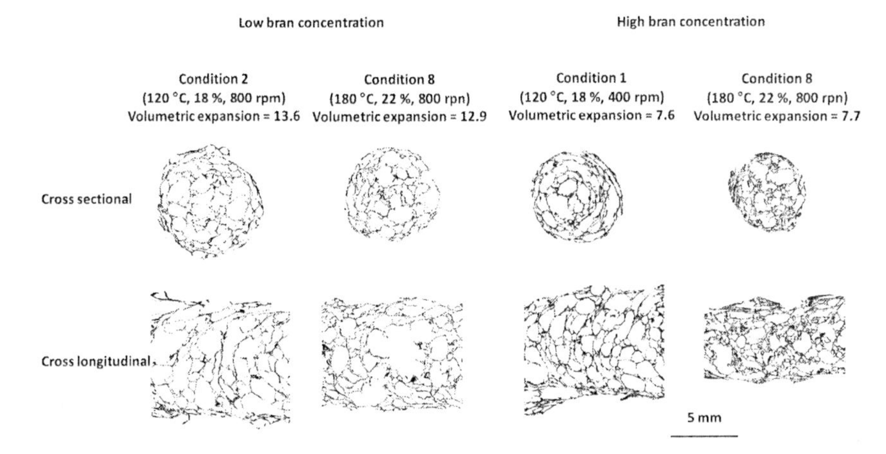

Fig. 6.8: Cross sectional and cross longitudinal X-ray tomography images of low and high bran concentration samples with similar volumetric expansion index

A similar relationship between the bulk expansion properties and the cellular structure to the one found for the refined flour could be established for bran-containing samples. This relationship can be observed when comparing the low bran sample extruded at condition 2 (120 °C, 18 % water content in the feed and 800 rpm) and at condition 8 (180 °C, 22 % water content in the feed and 800 rpm). Both had a similar volumetric expansion (VEI = 13.6 vs. 12.9) (see Fig. 6.3). Nevertheless, as it can seen in Fig. 6.8, the sample extruded at condition 8 showed a bulk expansion closer to anisotropy and a finer structure with a significant higher density (N_c = 2120 cm^{-3} vs. 800 cm^{-3}) of smaller cells (MCS = 410 μm vs. 540 μm) than the one extruded at condition 2. A similar trend can be reported between the high bran samples extruded at condition 1 (120 °C, 18 % water content in the feed and 400 rpm) and at condition 8 (180 °C, 22 % water content in the feed and 800 rpm). Both samples exhibit the same volumetric expansion (VEI = 7.6 vs. 7.7) (see Fig. 6.3). At condition 8, the longitudinal expansion was favored and the cellular structure of the resulting sample was finer with a significant lower density (N_c = 2830 cm^{-3} vs. 1610 cm^{-3}) of smaller cells (MCS = 230 μm vs. 380 μm) than the sample extruded at condition 1 (Fig. 6.8). The samples closer to isotropic expansion also showed cells being closer to spherical shapes (Fig. 6.8). For instance, the low bran samples (LB) extruded at condition 8 (LB8) in the longitudinal direction while at condition 2 (LB2) cells squeezed in the sectional section, closer to oblate ellipsoids could be visually observed (Fig. 6.7).

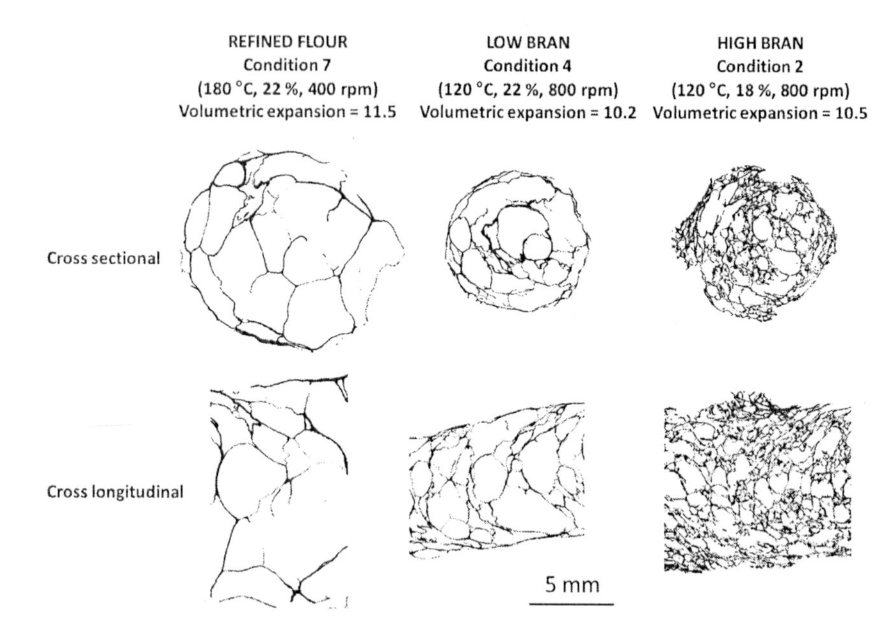

Fig. 6.9: Cross sectional and cross longitudinal X-ray tomography images of extruded samples with increasing bran concentration, having a similar volumetric expansion index

The effect of bran on the cellular structure was obtained when comparing samples with different bran concentration at similar volumetric expansion index (see Fig. 6.9). For this, the refined flour sample extruded at condition 7 (180 °C, 22 % water content in the feed and 400 rpm, VEI = 11.5), low bran sample extruded at condition 4 (120 °C, 22 % water content in the feed and 800 rpm, VEI = 10.2) and high bran sample extruded at condition 2 (120 °C, 18 % water content in the feed and 800 rpm, VEI = 10.5) were compared. Increasing the bran concentration in these samples favored longitudinal bulk expansion (LEI = 0.86 vs. 0.99 vs. 2.66, respectively for RF7, LB4 and HB2). It significantly increased the cell density (N_c = 50 cm^{-3} vs. 220 cm^{-3} vs. 3650 cm^{-3}, respectively). Furthermore the mean cell size (MCS = 1730 µm vs. 890 µm vs. 210 µm, respectively) and the mean cell wall thickness (MCWT = 75 µm vs. 55 µm vs. 30 µm, respectively) were significantly decreased (Fig. 6.9). More spherical cells were also generated when increasing the bran concentration (Fig. 6.9).

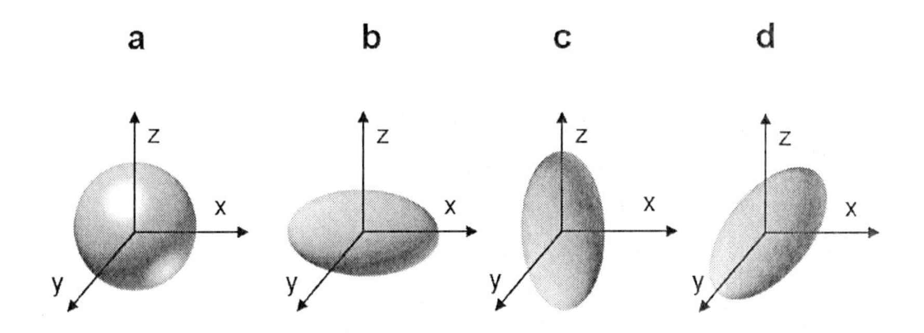

Fig. 6.10: Representation of bubble orientation during growth in the case of isotropic growth at the die exit (a), cell growth oriented in parallel to the die (b) or cell growth oriented in the sectional direction of the die (c and d) (reproduced from Robin et al., 2010a)

According to these results recipes with increasing bran content showed bulk expansion behavior closer to isotropy and finer structures with a higher density of small cells. On the contrary, refined flour samples showed favored sectional bulk expansion and coarser cellular structures. In case of a favored sectional expansion, cell closer to oblate ellipsoids would be expected (Fig. 6.10c and d). In the case of isotropic bulk expansion the bubbles would be expected to grow evenly in all directions. This would result in spherical cells in the final product (see Fig. 6.10a). With a favored longitudinal expansion, cells stretched into the direction of extrusion (x) would be expected (see Fig. 6.10b). Although some high bran samples showed favored longitudinal expansion, the orientation of the cells was difficult to appreciate due to the low size of the cells.

Increasing the bran concentration led to cell with shapes closer to spheres. The mean cell size (MCS) and the cell density (N_c) were obtained by filing the foam cells with virtual spheres. The relationship between the cell density and the mean cell size should follow a power law relationship according to equation (6.1):

$$N_c \propto MCS^{\alpha} \tag{6.1}$$

The relationship between the relative density (D) and the mean cell size should then be given by equation (6.2):

$$D = 1 - \frac{4}{3}\pi N_c \left(\frac{MCS}{2}\right)^{\alpha} \tag{6.2}$$

The power law index α would equal 3 in the case of cells being perfectly spherical. As shown in Fig. 6.11, the low bran and high bran samples showed an index α of 2.5 and 2.7, respectively. This value may confirm the shapes closer to spheres obtained with the samples containing bran. The refined flour samples appeared to form two groups that could be differentiated according to their cell density: RF2, 5, 6 & 7 having a higher number of cells than RF1, 3, 4 and 8 (Fig. 6.11). Due to the high scattering of these data, an overall trend could not be determined.

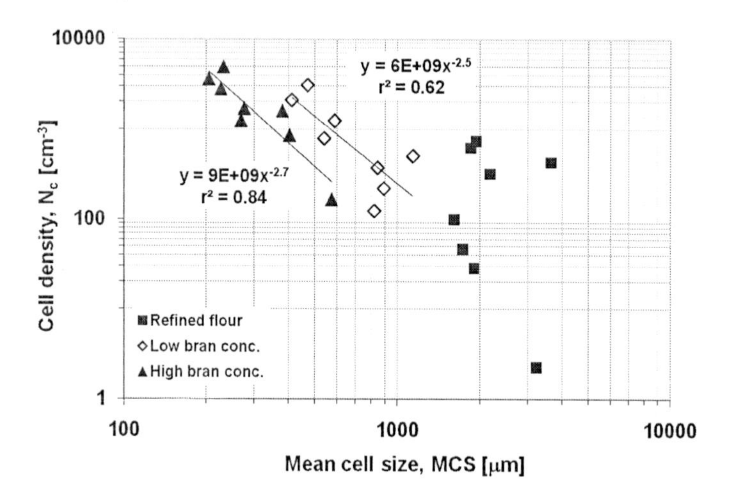

Fig. 6.11: Cell density N_c vs. mean cell size (MCS) of extruded samples at different process conditions

These results confirm that a relationship between the samples bulk expansion properties and their cellular characteristics exists. This is in good agreement with previous reports by Robin et al. (2010a), Della Valle et al. (1997) and Desrumaux et al. (1998).

VI.3.2. Effect of melt rheological properties on expansion properties

The rheological properties of the matrix surrounding the bubbles determine their resistance to growth (Moraru and Kokini, 2003). In Chapter V, the melt shear viscosity was measured with the rheometer in similar conditions of extrusion as with the die. The physicochemical properties of the ingredients were similar when using the rheometer or the die at same process conditions (see Chapter V.2.1.3). In this Chapter the rheology data used to explain the expansion properties of extruded samples with increasing bran concentration. The range of apparent shear rates achieved with the current rheometer (between 10 and 30 s^{-1}) was much lower than the theoretical ones at the die exit. Indeed an apparent shear rate of about 600 s^{-1} would be expected with the die. In order to explain the expansion data with the rheology data, the shear viscosity should be calculated at the same shear rate as obtained at the die. To obtain the shear viscosity at the shear rate in the die, extrapolation of the shear viscosity values measured with the rheometer would be necessary. In this study, it was preferred to focus and discuss on the trends between the shear viscosity and the expansion properties.

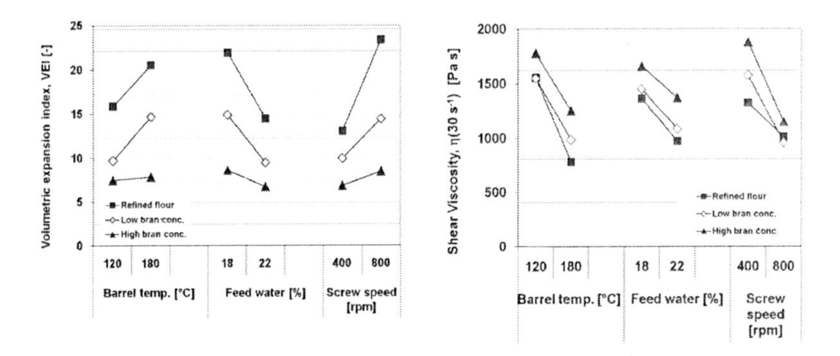

Fig. 6.12: Effects of process parameters on the volumetric expansion index and shear viscosity (at 30 s^{-1}) (the distance between two grid lines represents the least significant difference)

The correlation between the shear viscosity and the expansion data index was not satisfying (see in Appendices in Tab. 10.5 and 10.6). The relationship was not improved when adjusting the shear viscosity data according to the saturated vapor pressure, at the melt temperature, as suggested by Bouzaza et al. (1996) (see Chapter II and equation 2.15). This means that neither the shear viscosity nor the pressure were the sole driver of expansion, within the tested conditions. The results of the statistical analysis shows the effect of extrusion parameters on volumetric expansion index (VEI) and shear viscosity at a shear rate of 30 s^{-1} (Fig. 6.12). Increasing the screw speed and the barrel temperature increased the volumetric expansion (VEI). The volumetric expansion index was decreased with the water content. The effect of the process conditions on VEI was reduced when increasing the bran concentration. It was not significant at high bran concentration (Fig. 6.12). This is likely caused by the difference in mechanical properties between bran and starch.

When increasing the water content, a significant decrease in both volumetric expansion and shear viscosity of the refined flour and high bran samples can be observed (Fig. 6.12). This is consistent with previous reports (e.g. Guy and Horn, 1988; Della Valle et al., 1997). This decrease in volumetric expansion can be attributed to the increased shrinkage of the extrudate after maximum expansion (see shrinkage ratio in Fig. 6.13). This shrinkage is due to the reduced melt viscosity. On the opposite, a significant increase in volumetric expansion and a significant decrease in shear viscosity can be observed for the refined flour and low bran samples when increasing the barrel temperature and screw speed (Fig.

6.12). The decrease in temperature was favorable for volumetric expansion. The decrease in shear viscosity with the screw speed favored volumetric expansion (Fig. 6.12).

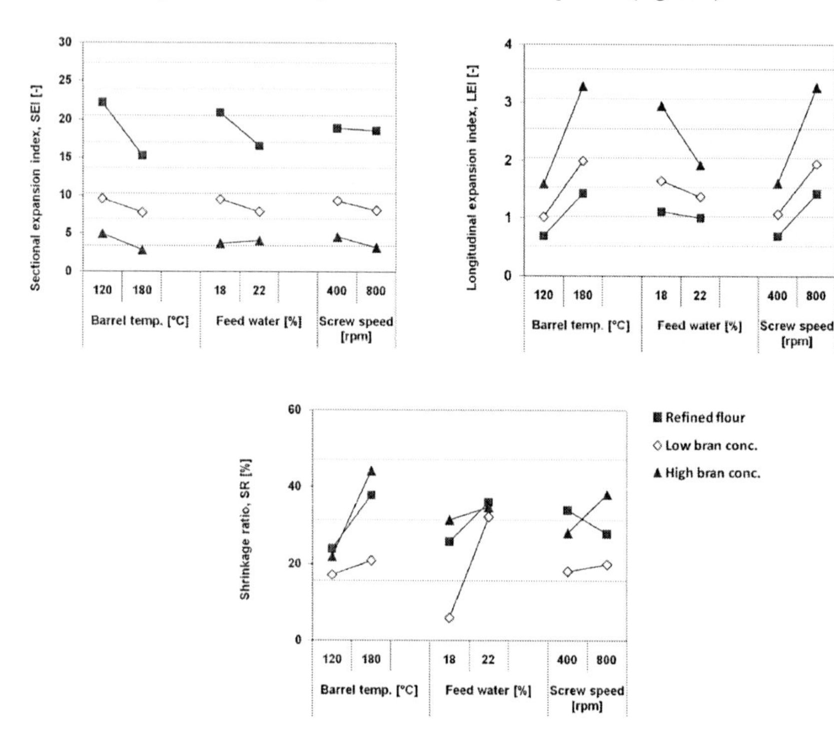

Fig. 6.13: Effects of process parameters on sectional expansion index, longitudinal expansion index and shrinkage ratio (the distance between two grid lines represents the least significant difference)

The sectional expansion (SEI) was only significantly decreased for the refined flour when increasing barrel temperature and water content as shown in Fig. 6.13. Therefore, the change in volumetric expansion of the extruded samples with the process parameters was mainly attributed to the change in longitudinal expansion (Fig. 6.13). Both the longitudinal (LEI) and volumetric expansion (VEI) showed the same trends with the shear viscosity when modulating the process parameters (Fig. 6.12 and 6.13). A positive trend was reported between the shear viscosity and the sectional expansion index (SEI) of the refined flour when decreasing the barrel temperature and the water content in the feed (see Fig. 6.12 and Fig. 6.13).

The shear viscosity may not be the sole driver of cell growth. For instance refined flour samples extruded at condition 7 (180°C, water content in the feed of 22 % and 400 rpm) and 8 (180°C, water content in the feed of 22 % and 800 rpm) exhibited similar shear viscosity (670 MPa vs. 610 MPa). Nevertheless, the sample extruded at condition 7 exhibited a higher maximum sectional expansion index (SEI_{max}) (28.6 vs. 13.4). This difference in maximum sectional expansion at the die may be due to difference in extensional viscosity/elastic properties of the melt.

Increasing bran concentration decreased volumetric expansion index. The change in viscoelastic properties of the melt and/or an increased collapse of the sample during growth when adding bran may explain the reduced expansion volume. At similar extrusion conditions, the maximum sectional expansion index (SEI_{max}) of the bran-containing samples was lower than the sectional expansion index (SEI) of the refined flour (see Appendices Tab. 10.6). Therefore, this excludes collapse of the structure being the major reason for the lower sectional expansion (SEI) at increasing bran content.

As shown in Fig. 6.12, the shear viscosity was significantly increased only at the highest bran concentration. The effect was nevertheless process condition-dependant. It was only significant at the highest barrel temperature, screw speed or lowest water content (Fig. 6.12). The increase in shear viscosity with the high bran concentration may hinder the bubble growth and reduce volumetric expansion. The refined flour and low bran concentration samples showed no significant difference in shear viscosity (Fig. 6.12). Nevertheless, the maximum sectional expansion index (SEI_{max}) of the low bran concentration samples were reduced compared to those of the refined flour samples (RF) (see in Appendices in Tab. 10.6). This may partially be explained by the extensional properties of the melt. The role of the extensional viscosity on expansion was recently highlighted by Pai et al. (2009). The authors reported a decrease in expansion properties of corn flour when adding corn bran. They linked this loss in expansion to an increase in extensional and shear viscosities, measured with off-line techniques. From measurement of the entrance pressure drop (see Chapter V.5.3), it is difficult to conclude on the effect of bran concentration on the melt extensional viscosity. Nevertheless, it is likely that the degree of elongation of the melt is reduced when increasing the bran concentration. This may be due to low adhesion between starch and bran. This may therefore reduce the capacity of the bubbles to expand. According to the theory of Bouzaza et al. (1996), the decrease in elastic recovery at the die exit may also explain the increased longitudinal expansion (LEI) at the expense of the sectional expansion (SEI) observed when increasing the bran concentration. Additionally, according to the classical nucleation theories, adding nucleation agents (such as bran) would decrease the free energy barrier to nucleate bubbles (Lee, 2000). A lower degree of supersaturation would be required for nucleating new bubbles and this would lead to earlier nucleation in the die and favored longitudinal expansion due to stretching of the bubbles in the direction of the melt flow (Lee, 2000; Della Valle et al., 1997).

The rheological properties were assessed in the conditions of extrusion in the die. Nevertheless, once leaving the die the melt undergoes evaporating cooling. Its rheological properties are then changed according to the kinetics of cooling and water evaporation. Therefore, the time necessary for the structure to reach the critical viscosity at which the bubble stops to growth will depend on this kinetics. There are debates concerning the value of this critical viscosity. Nevertheless it is often referred in the literature at $T_g + 30$ K (Fan et al., 1994). At same water content and melt temperature, both the kinetics of heat and water transfer from the melt to the atmosphere will depend on the material properties and the structural characteristics of the expanding product (volume, porosity, number of bubbles, their connectivity and size, thickness of the bubble walls and surface porosity) (Gibson and Ashby, 1997). These properties will differ according to the sample composition. Increasing the bran concentration reduced volumetric (hence decreased the porosity) expansions (Fig. 6.12) and mean cell wall thickness and increased the number of small cells (see in Appendices in Tab. 10.6). The samples with increased bran concentration also showed higher surface porosity which indicates rapid vapor loss.

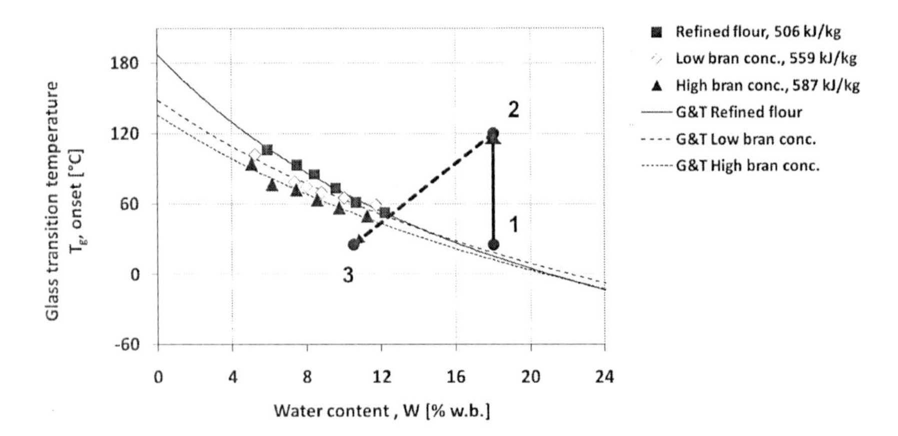

Fig. 6.15: Hydrothermal path of samples with increasing bran concentration extruded at condition 1 (120 °C, water content in the feed of 18 % and 400 rpm) (Step 1 to 2 corresponds to heating in the extruder, step 2 to 3 corresponds to cooling and drying of the melt at the die exit)

Fig. 6.15 shows the hydrothermal path ingredients extruded at condition 1 (120 °C, water content in the feed of 18 % and 400 rpm). After hydration of the powder in the extruder to 18 % (stage 1), the mix powder-water is heated to 120 °C (stage 2). During heating in the extrusion starch is melted and mainly amlyopectin molecules are broke down (see Chapter IV.2.6). Wheat bran particles are alos broken down depending on the extrusion conditions (see Chapter IV.3.2). When the melt is leaving the die, it cools down and water is evaporating. Its temperature decreases below the glass transition temperature. Straight after extrusion the water content in the samples was about 10.5 % at room temperature (stage 3). As seen in Fig. 6.15, the glass transition temperature of starch is decreased when increasing the bran concentration. All other conditions being the same, this may increase the time necessary to reach the critical viscosity at which the bubble growth ceases. This may therefore allow bubbles to grow for longer and increase their final dimensions.

X-Ray tomography pictures of the surface of the samples extruded at same process conditions (condition 1, 120 °C, water content in the feed of 18 % and 400 rpm) showed an increased surface porosity with increasing bran concentration (Fig. 6.16). This can be attributed to the premature rupture of the bubbles during growth as earlier reported by Guy and Horne (1988) and Yanniotis et al. (2007). The premature rupture of the bubbles can reduce significantly the growth of the bubbles and their final size. The rupture of bubbles is due to the low compatibility between the bran particles and the continuous starch phase. Therefore, increasing the bran volume fraction increases the number of points of rupture across the bubble membrane. Additionally, the rupture is favored at a critical bubble membrane thickness close to the bran particle size. As earlier reported the bran particles dimensions were also modified during extrusion depending on the extrusion conditions (see Chapter IV). This may also influence the critical membrane thickness at which it is ruptured. At same process conditions, increasing the bran concentration significantly decreased the mean cell wall thickness (see in Appendices in Tab. 10.6), likely further favoring the rupture of the bubbles.

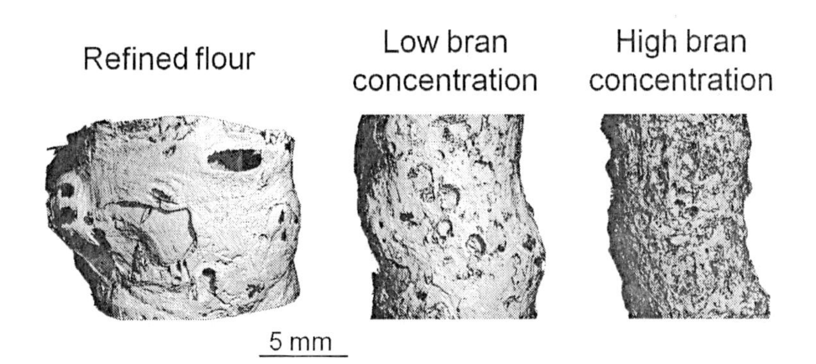

Fig. 6.16: 3D X-ray tomography pictures of samples made of refined wheat flour, low bran or high bran concentration and extruded at condition 1 (120 °C, water content in the feed of 18 % and 400 rpm)

Burst and shrinkage of the bubbles may be considered to be close phenomena. Both lead to a significant reduction in the bubble size. They can be differentiated according to their duration. Indeed, as observed visually, burst of the bubbles when increasing the bran concentration occurred in a very short time after the exit of the die. Shrinkage of the bubbles could be observed over a longer period of time after exit of the die. Shrinkage of the extruded samples after expansion at the die occurred as shown in Fig. 6.13. The shrinkage ratio was only significant increased for the refined flour samples extruded at different temperature and for the low bran samples extruded at different water content (Fig. 6.13). This can be related to a decrease in shear viscosity (see Fig. 6.12). The effect of the screw speed was not significant. The trends were not clear depending on the bran concentration. Increasing the bran content to the highest level appeared to lead to a similar shrinkage ratio (SR) to the refined flour when varying barrel temperature and water content (Fig. 6.13). The samples extruded at the intermediate level of bran showed a lower shrinkage ratio (SR) while at higher bran content the shrinkage was similar to the one of the refined flour. As previously mentioned increasing the bran concentration increases the shear viscosity (Fig. 6.12). A higher shear viscosity should be less favorable for shrinkage. Indeed a higher viscosity of the membrane around the cell enables to withstand the pressure inside the bubbles for a longer time. Nevertheless, this may be counteracted by the decrease in the glass transition temperature of starch with increasing the bran concentration (see Chapter IV). This means a lower viscosity of the starch phase. At constant melt temperature, this lower glass transition temperature may also mean a longer time to reach the critical viscosity at which the bubble ceases to grow and therefore longer time for shrinkage.

VI.4. Conclusions

Increasing the bran concentration increases the longitudinal expansion and the bulk density of extruded foams at the expense of the sectional and volumetric expansion. When adding bran the expansion is favored in the longitudinal direction (parallel to the die). The addition of bran generates fine cellular structures having a high number of small cells. The effect of bran on the expansion properties can be explained by investigating its effect on each step of the expansion mechanism. Increasing the bran concentration increases the density of cells in the final product. This can be attributed to a higher degree of nucleation due to the presence of non wetted insoluble bran particles. The presence of bran likely induces heterogeneous nucleation. This increase in the degree of nucleation with bran may affect the rheological properties of the melt and bubble growth. This can be caused by the formation of thinner bubble walls or increased shear at the bubble interfaces.

In order to grow in the melt the nucleated bubbles have to overcome the resistance created by the shear viscosity. Under the tested conditions, the shear viscosity of the melt is significantly increased only at the highest bran concentration. Furthermore, this increase is process condition-dependant. The increase in shear viscosity at the highest bran concentration may hinder expansion and explain the reduced maximum sectional expansion at the die. At low bran concentration the expansion properties are already reduced but the shear viscosity of the melt is not significantly changed. This may be explained by a difference in extensional viscosity when increasing bran concentration.

Shrinkage of the expanded samples was measured after the maximum expansion at the die exit. The hypothesis that a higher collapse of the extrudate during expansion is responsible for the lower final sectional expansion of samples containing bran can be excluded. A higher surface porosity can be observed when increasing the bran concentration. This may be the result of the rupture of the bubbles during growth. This can be due to the low adhesion properties at the interface between the continuous starch phase and the bran particles. This reduces the capacity of the cells to elongate due to their premature rupture and participate to the reduced bubble growth. This can also explain the reduced expansion volumes of bran-containing samples. The combination of an increased degree of nucleation, increase in shear viscosity, decrease in melt elastic properties, premature rupture of the bubbles and shrinkage of the structure may explain the lower sectional expansion of starchy melts enriched in wheat bran. As all these parameters are interdependent it is nevertheless difficult to evaluate their individual effect on expansion.

CHAPTER VII. MECHANICAL PROPERTIES OF EXTRUDED FOAMS

VII.1. Introduction and hypothesis

The mechanical properties of solid foams are driven by their bulk dimension and shape, the distribution of the continuous phase (cell walls) and the dispersed phase (porosity) and the physicochemical properties of the material (mentioned hereafter as cell wall material). All these properties are resulting from expansion of the melt. The properties of cell walls and porosity govern the product appearance and textural properties. The density of a product is also of importance as products are very often commercialized by weight. At similar bulk properties, the textural properties of a product may be varied depending on the microstructure of the samples. Defining the optimal cellular structure characteristics of samples containing wheat bran may enable to improve their textural properties. In order to analyze the effect of the cellular structure characteristics obtained when incorporating wheat bran on the mechanical properties a three point bending test was used. The instrumental measurements obtained from this test can be linked to in-mouth sensory properties such as crispiness (see Chapter II.6.1). From this test, the stress at rupture and elastic modulus were extracted. Both enable to normalize the force necessary to break the samples depending on their diameter. Therefore, the effect of the porosity on the mechanical properties can be determined, regardless of the sample diameter.

In Chapter V the effect the bran concentration on bulk expansion properties and cellular structure was discussed. These data are used in this Chapter to analyze the effect of the bulk expansion and cellular structure characteristics on the mechanical properties of the foams. Then the effect of the cellular structure on the mechanical properties of foams with increasing bran concentration is reported and discussed. The cell wall material properties may also have a significant effect on the mechanical properties of extruded foams. These properties depend on the composition and on ingredient transformations during extrusion. They also depend on the water content of the cell walls. The mechanical properties of the cell wall material were obtained by thermomolding and application of large amplitude deformations.

The effect of bran concentration on the mechanical properties of both the cell wall material and extruded foams is reported in this Chapter.

VII.2. Mechanical properties of cell walls

VII.2.1. Effect of bran concentration on mechanical properties of cell walls

The stress at rupture σ_s and elastic modulus E_s of the cell wall material were obtained by thermomolding of the extruded foams. The conditions were chosen in order to modify the least the physicochemical properties of the material during sample preparation. One hypothesis was that both the extrusion conditions and bran concentration may have an effect on the mechanical properties of the cells walls. To investigate the effect of processing on cell walls, the mechanical properties of extruded samples with two mechanical energy (SME) input were investigated: at condition 1 (120 °C, 18 % water content in the feed and 400 rpm) with a range of specific mechanical energy from 500 to 600 kJ kg^{-1} and at condition 8 (180 °C, 22 % water content in the feed and 800 rpm) with a range of specific mechanical energy from 280 to 415 kJ kg^{-1}.

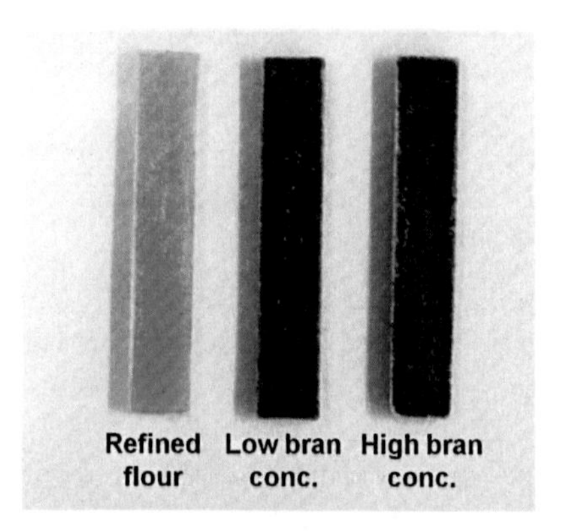

Fig. 7.1: Thermomolded bars of samples extruded at condition 8 (180 °C, water content in the feed of 22 % and 800 rpm)

The appearance of the bars obtained after thermomolding is shown in Fig. 7.1. The bar made of extruded refined flour was transparent indicating that no porosity remained. The stress at rupture σ_s and elastic modulus E_s of the tested bars equilibrated at a water activity a_w of 0.33 are shown in Tab. 7.1. The stress at rupture σ_s of the samples extruded at condition 1 (120 °C, 18 % water content in the feed and 400 rpm) was only significantly reduced when increasing the bran concentration to the highest level (Tab. 7.1). The elastic modulus E_s of these samples was significantly reduced by the increasing content of wheat bran (Tab. 7.1). The samples obtained at condition 1 (120 °C, 18 % water content in the feed and 400 rpm) were produced at a significant higher specific mechanical energy input (SME) compared to the samples produced at condition 8 (180 °C, 22 % water content and 800 rpm) (Table 1). A higher specific mechanical energy results in a higher degree of transformation of starch (see Chapter IV.2.1). Despite this difference in process conditions, no significant differences in mechanical properties could be observed (Tab. 7.1).

Tab. 7.1: Stress at rupture σ_s and elastic modulus E_s of thermomolded samples extruded at condition 1 (120 °C, water content in the feed of 18 % and 400 rpm) and at condition 8 (120 °C, water content in the feed of 22 % and 800 rpm) and equilibrated at water activity of a_w 0.30 (specific mechanical energy, SME, is indicated)

	Condition 1 (SME ≈ 550 kJ kg^{-1})		Condition 8 (SME ≈ 450 kJ kg^{-1})	
	σ_s [MPa]	E_s [MPa]	σ_s [10^3 MPa]	E_s [10^3 MPa]
Refined flour	27.5 ± 3.4	3.00 ± 0.62	29.0 ± 4.4	2.65 ± 0.53
Low bran conc.	24.5 ± 3.0	2.01 ± 0.15	23.6 ± 1.7	1.93 ± 0.22
High bran conc.	18.0 ± 2.5	1.55 ± 0.25	19.0 ± 2.7	1.62 ± 0.34

Fibers are the main component of wheat bran particles. They are mostly insoluble and only slightly solubilised during extrusion (see Chapter IV.3). Due to their low compatibility with the continuous starch network, bran particles can be considered as filler of the continuous starch matrix. As reported for plastics reinforced with synthetic fibers, the properties of the fillers influence the mechanical properties of the composite material (Xanthos, 2005). These properties of the fillers include their mechanical properties, phase continuity, volume fraction, particle size and particle size distribution, aspect ratio (ratio of the largest dimension to the smallest) and orientation (Xanthos, 2005). Fillers also induce anisotropy in the mechanical properties depending on the direction of the applied force. In the case of fibers showing discontinuity in their phase distribution (as observed for the bran particles), the shear stress applied to the composite matrix would build-up at the interface between the two phases and the rupture occurs either by fibers fracture or adhesion fracture at the interface (Xanthos, 2005). The rupture at the interface between the bran particles and the continuous starch phase is likely to be favored when the bran fraction is increased. It may create a fracture path occurring mainly at the interface between the bran particles and the continuous starch phase. This may explain the observed decrease in both stress modulus at rupture and elastic modulus.

The bran-containing samples extruded condition 1 and 8 were produced with two different bran particle sizes: "fine and coarse" and two different specific mechanical energy values (Chapter III, Tab, 3.2). The bran particle sizes are changed when increasing the specific mechanical energy input as shown in Chapter IV.3.2. At same bran volume fraction, decreasing the particle dimension caused by an increase in mechanical shear in the extruder means that the density of the particles is increased. The bran particles tended to be homogenously distributed and oriented in the melt flow direction during cell growth (no difference in orientation could be observed depending on radial or longitudinal direction) (see in Chapter IV.3.2). According to Osswald and Menges (1995) and at same volume fraction of dispersed particles, the maximum strength of a composite matrix containing dispersed fiber particles would be obtained when the fibres are oriented in the direction of the applied stress. An aspect ratio higher than 1 and fibers oriented perpendicular to the applied force would show a lower maximum strength compared to fibers with an aspect ratio of 1. At same volume fraction and aspect ratio of fibers, having a higher density of small dispersed particles would increase the strength of the composite matrix (Rothon, 2003). A similar approach can be taken to the protein phase in the refined wheat flour. As observed in Chapter IV.3.2 in Fig. 4.17a, mainly large spherical protein spots (stained in green) of about 10-15 μm diameter can be observed at low specific mechanical energy (condition 7, 180 °C, 22 % water content in the feed and 400 rpm, SME = 233 kJ kg^{-1}). At high specific mechanical energy (condition 2, 120 °C, 18 % water content in the feed and 800 rpm, SME = 607 kJ kg^{-1}) mainly linear protein structures were present (Fig. 4.17b). The morphology of the protein phase may also affect the mechanical properties of the samples. The protein/starch interface properties were also reported to influence the mechanical properties of extruded cereals (Chanvrier et al., 2006). As reported earlier, there was no significant difference in the mechanical properties of the extruded cell wall material depending on the process conditions. Therefore the difference in particle size between the two bran qualities and/or the reduction in bran particle size during extrusion did not induce significant differences.

The mechanical properties of the bran particles on their own were not assessed. Several authors measured the mechanical properties of wheat bran layers using dynamic mechanical thermal analysis (Antoine et al., 2003; Grefeuille et al., 2007; Hemery et al., 2010). They reported that bran exhibits an elastoplastic rheological behaviour. The degree in extensibility (elastic modulus) and stiffness (strain) of bran is different through the bran layers depending on the layer composition (Antoine et al., 2003) as well as water content and temperature (Hemery et al., 2010). Although the conditions of water content used in the previously cited works differed from those used in this study, wheat bran may show a lower stress at rupture and elastic properties than the continuous phase and also contribute to the reduced values of stress at rupture and elastic modulus of the composite matrix when increasing the bran volume fraction.

VII.2.2. Effect of water distribution on mechanical properties of cell walls

Additionally to the effect of dispersed bran particles, competition for water between fibers, present in bran, and starch may affect the mechanical properties of the composite matrix. At extrusion condition 1 (120 °C, 18 % water content in the feed and 400 rpm) and water activity of 0.33, an increase of bran concentration is coupled to a reduction in water content (Tab. 7.2). The water content significantly drops from 8.40 % to 8.00 % and 7.44 % for the low bran and high bran concentration samples, respectively. This effect can be attributed to the lower hygrocapacity of the bran fibres as reported earlier in Chapter IV2.5.1.1. The glass transition temperature of starch was reduced when increasing the bran concentration at same water activity (Tab. 7.2). This is attributed to more free water available for the starch phase (see Chapter IV.2.5.3.2). This higher available free water for the starch phase leads to a decrease in its glass transition temperature (see Chapter IV.2.5.3.2). At a water activity of 0.33, the decrease in glass transition temperature between the refined flour and high bran concentration samples was 13.2 K (Tab. 7.2). The same trend was found at condition 8. At this extrusion condition and at a water activity of 0.33, the decrease in glass transition temperature between the refined flour and high bran concentration was about 9.0 K (Tab. 7.2). Therefore, at room temperature T and same water activity, increasing the bran concentration induces a decrease in the T-T_g difference of 9.0 or 13.2 K depending on the extrusion condition (Tab. 7.2). The decrease in the starch glass transition with the bran concentration may influence the stress at rupture and elastic modulus of the bran-containing matrices.

Tab. 7.2: Water content and glass transition temperature $T_{g,\ onset}$ at of samples iwht increasing bran concentration extruded at condition 1 (120 °C, water content in the feed of 18 % and 400 rpm) or condition 8 (180 °C, water content in the feed of 22 % and 800 rpm) and equilibrated at a_w 0.33 (specific mechanical energy, SME, is indicated next to the extrusion condition number)

	Condition 1 (SME ≈ 550 kJ kg^{-1})				Condition 8 (SME ≈ 450 kJ kg^{-1})		
	Refined flour	Low bran conc.	High bran conc.		Refined flour	Low bran conc.	High bran conc.
Water content [% w.b.]	8.40 ± 0.01	8.00 ± 0.00	7.44 ± 0.00		8.45 ± 0.00	8.27 ± 0.00	7.45 ± 0.02
$T_{g,\ onset}$ [°C]	85.0 ± 1.4	74.7 ± 0.9	71.8 ± 0.1		80.0 ± 4.2	73.3 ± 0.4	71.0 ± 0.0

VII.3. Mechanical properties of extruded foams

VII.3.1. Application of the Gibson-Ashby model to mechanical properties of extruded foams

The mechanical properties of the extruded foams depending on extrusion condition and bran concentration were determined using a three point bending test.

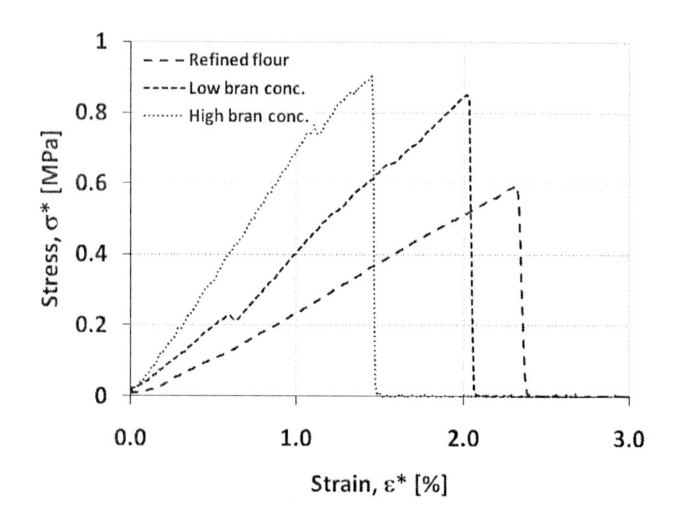

Fig. 7.2: Compression curves stress σ^* vs. strain ε^* of samples with increasing bran concentration extruded at condition 8 (180 °C, water content in the feed of 22 % and 800 rpm) and equilibrated at a water activity a_w of 0.30

Typical stress-strain curves obtained after bending of the extruded samples with increasing bran concentration are displayed in Fig. 7.2. The compression curve shows a linear increase in stress at low strain and a fracture at higher strain, resulting in a marked decrease in stress. Such curves are typical of elastic brittle foams as described by Gibson and Ashby (1997) (see Chapter II, Fig. 2.11). The force F necessary to rupture the samples depending on bran concentration and process conditions is shown in Fig. 7.3. This force was decreased when increasing the bran concentration. A similar trend was reported by Moore et al. (1990), Wang et al. (1993), Onwulata et al. (2001) or Brennan et al. (2008). The correlation between the diameter of the samples (obtained by calculation from the sectional expansion index) and the breaking force was poor (see in Appendices in Tab. 10.6). This indicates that the sample diameter was not the sole driver of the breaking force. Other factors such as the porosity, cellular structure characteristics and cell wall material properties may also affect the breaking force.

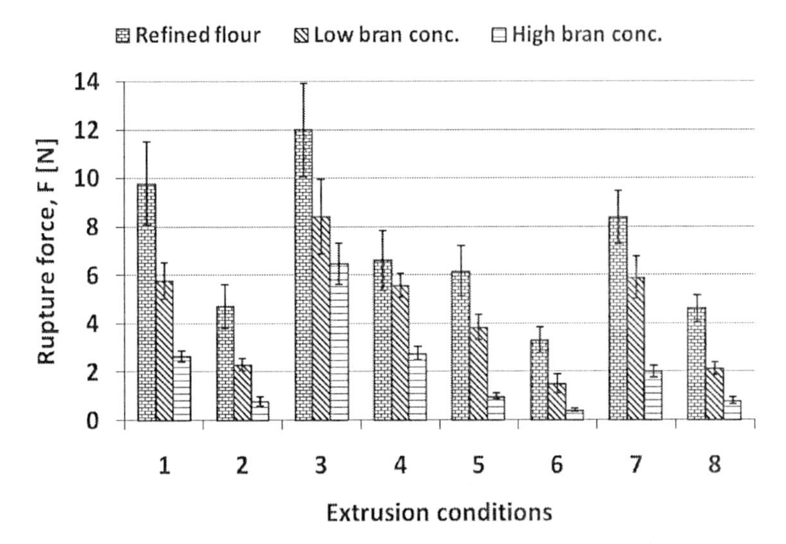

Fig. 7.3: Effect of bran concentration on the rupture force F of samples extruded at same conditions (see in Chapter III in Tab. 3.2 for extrusion conditions) and equilibrated at a water activity a_w of 0.30

In order to normalize the effect of the diameter of the samples on the mechanical properties, the stress at rupture and elastic modulus were calculated (see equations 3.23 and 3.25). Stresses at rupture σ^* from 0.27 MPa to 0.87 MPa, 0.47 MPa to 1.80 MPa and 0.55 MPa to 1.48 MPa were obtained for the refined flour, low bran and high bran concentration samples, respectively (see in Appendices in Tab. 10.6). This was within the range of values reported by Lourdin et al. (1995) for flat expanded starches or by Robin et al. (2010a) for flat expanded wheat flour. The range of elastic modulus E^* was 0.33 - 3.53 MPa, 1.25 - 6.60 MPa and 1.68 - 14.06 MPa for the refined flour, low bran and high bran concentration samples, respectively. The standard deviation of these values ranged from 11 % to 32 %. This high standard deviation reflects the high variability in the sample surface and internal structure, especially of those made of pure refined flour.

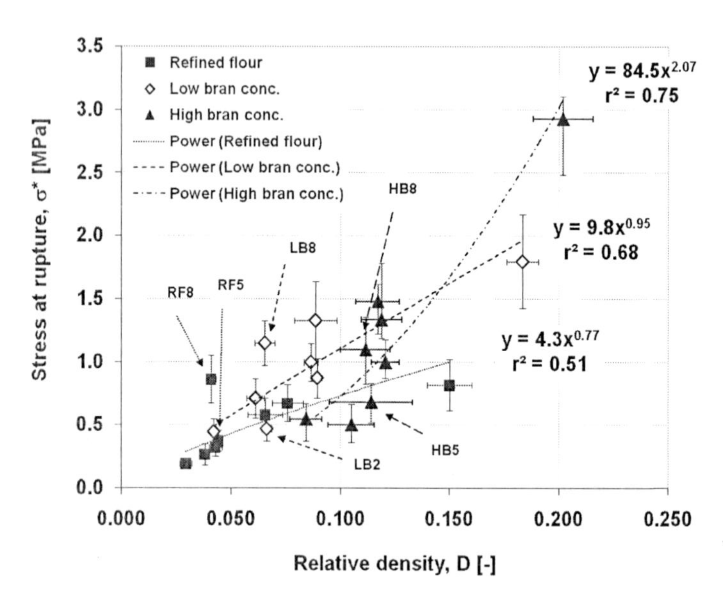

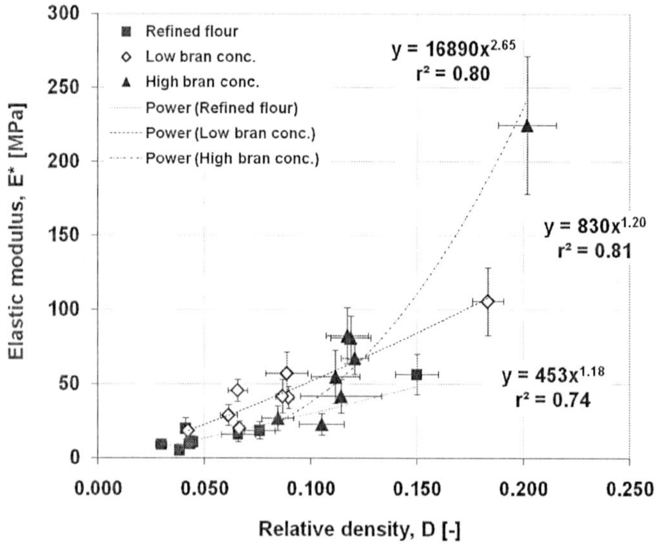

Fig. 7.4: **Stress at rupture σ^* and elastic modulus E^* vs. relative density D of extruded samples with increasing bran concentration extruded at different conditions and equilibrated at a water activity a_w of 0.30**

Fig. 7.4 shows the stress at rupture and elastic modulus of samples with increasing bran concentration extruded at different conditions depending on their relative density. The stress at rupture and the elastic modulus of the extruded foams increased with their relative density (Fig.7.4). The relative densities obtained in this work were within the range where the analogy to solid foams can be made (usually valid for $D < 0.030$) (Gibson and Ashby, 1997). The Gibson-Ashby model is linking the mechanical properties of solid foams to their relative density according to a power law relationship (equations 2.25 and 2.26). This model was applied to the measured values of stress at rupture and elastic modulus according to the bran concentration (Fig. 7.4). The correlation factor r^2 between the experimental values of stress at rupture and the power law fitting of these values according to the Gibson-Ashby model was not satisfying (Fig. 7.4). The correlation factor was nevertheless improved when increasing the bran concentration with respectively $r^2 = 0.51$, 0.68 and 0.75 for the refined flour (and bran-containing samples (Fig. 7.4). The correlation factor r^2 between the elastic modulus and the power law fitting of these values was satisfactory with respectively for the three recipes with increasing bran concentration; $r^2 =$ 0.74, 0.80 and 0.81 (Fig. 7.4). A similar power law index n was found for the refined flour and low bran concentration samples (n = 0.77 and 0.97). At higher bran concentration a higher power law index value was reported (n = 2.07) (Fig. 7.4). A similar observation can be made for the power law index m. Both the refined flour and low bran concentration samples exhibited similar m values (m = 1.18 and 1.20). At higher bran concentration a higher index m was found (m = 2.65) (Fig. 7.4). The power law indices n (0.77) and m (1.18) for the refined flour were similar to those reported by Lourdin et al. (1995) for flat expanded starches, which were 1.07 and 1.16, respectively. The power law indices n and m were nevertheless lower than the ones reported by Robin et al. (2010a) for flat expanded wheat flour supplemented with sucrose or sodium bicarbonate (n = 2.3, m = 3.7). In both these studies the range of relative densities was higher with a relative density (D up to 0.450). Such a wide range of relative density enables a better evaluation of the mechanical properties and a more precise fitting of the experimental values using power laws. The power law indices n and m for the refined flour and low bran concentration (were closer to the theoretical values obtained for open cell structures (n = 1.5 and m = 2) (Gibson and Ashby, 1997). The power law indices for the high bran recipe were closer to the theoretical values obtained for closed cells (n = 2 and m = 3) (Gibson and Ashby, 1997). Conflicting results about the degree of cell connectivity of extruded starchy foams have been reported (Warburton and Donald, 1992; Trater et al., 2005; Babin et al., 2007; Robin et al., 2010a). The results differed depending on the material (starch or flour), process conditions and measurement tools. The mechanical properties found for the refined wheat flour recipe which are closer to the ones of open-cell structures, appear in good agreement with the results reported by Babin et al. (2007). Nevertheless, they are contradicting the results of Warburton and Donald (1992), Trater et al. (2005) or Robin et al. (2010a). The later reported mechanical properties closer to those of closed-cell structures. The connectivity index I_{co} obtained in this study was close to 99 %, irrespective of the recipe and process conditions. This means that 99 % of the void volume was occupied by a connected cavity

(Babin et al., 2007). This would support the presence of an open-cell structure for the three recipes. However, cell walls thinner than the tomography detection limit may lead to errors in the determination of I_{co}. The performed fitting results using the Gibson-Ashby model may thus be questionable. Nevertheless, it clearly appears that the mechanical properties of the high bran recipe were significantly different from those of containing less or no bran.

VII.3.2. Effect of cellular structure on mechanical properties

VII.3.2.1. Effect of cellular structure on the mechanical properties at same bran concentration

As previously mentioned, the fitting of the experimental values with the Gibson-Ashby model was not satisfying. Within the sample population at a given product composition, discrepancies between the Gibson-Ashby model and the experimental values of shear stress could be observed (Fig. 7.4).

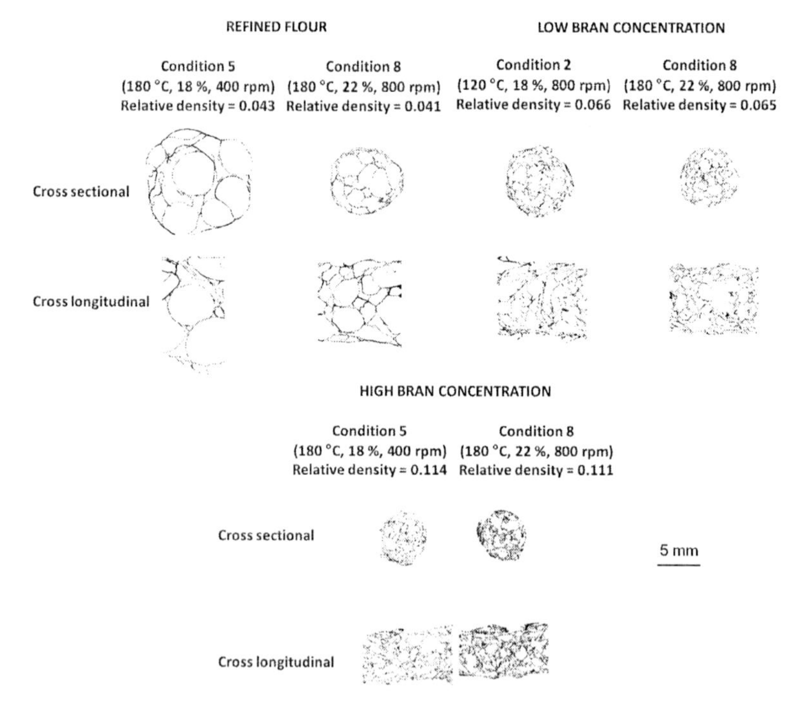

Fig. 7.5: **Cross sectional and cross longitudinal X-ray tomography images of extruded samples with increasing bran concentration at same relative density (*D*)**

Samples having close or same relative densities D show different mechanical properties. For instance, the refined wheat flour samples extruded at condition 5 (180 °C, 18 % water content in the feed and 400 rpm) and at condition 8 (180 °C, 22 % water content in the feed and 800 rpm) had same relative densities (D = 0.043 vs. 0.041) (Fig. 7.4 and in Appendices in Tab. 10.6). However the stress at rupture of the sample extruded at condition 8 was significantly higher (σ^* = 0.86 MPa vs. 0.32 MPa). This difference may be related to their difference in cellular structure. Indeed, as shown in Fig. 7.5, the sample extruded at condition 8 was characterized by a significant higher density of cells (Nc = 740 cm^{-3} vs. 320 cm^{-3}), smaller cells (MCS = 1940 µm vs. 2180 µm) and a lower mean cell wall thickness (MCWT = 30 µm vs. 45 µm) (see Appendices in Tab. 10.6). A similar observation could be reported for the low bran concentration foams extruded at condition 2 (120 °C, 18 % water content in the feed and 800 rpm) and at condition 8 (180 °C, 22 % water content in the feed and 800 rpm). Both sample showed same relative densities but the samples extruded at condition 8 had a higher stress at rupture and a finer cellular structure (see Fig. 7.5 and Appendices in Tab. 10.6). A similar trend was also observed when comparing the high bran concentration foams obtained at condition 5 (180 °C, 18 % water content in the feed and 400 rpm) and at condition 8 (180 °C, 22 % water content in the feed and 800 rpm). Both samples exhibited same relative densities but the samples extruded at condition 8 had a higher strength and a finer structure (see Fig. 7.5 and Appendices in Tab. 10.6). These results suggest that, irrespective of the bran concentration, finer structures with a higher number of small cells required more strength to be ruptured. This was in good agreement with previous reported studies (Babin et al., 2007; Robin et al., 2010). Additionally to their difference in cellular structure, samples extruded at different conditions of specific mechanical energy (SME) may also exhibit different cell wall material properties due to their different degree in starch transformation and/or change in bran/protein particle morphology. This may also explain their difference in mechanical properties. Nevertheless, as reported in Chapter VII.2, no significant change in mechanical properties was measured between samples extruded at two different conditions of specific mechanical energy.

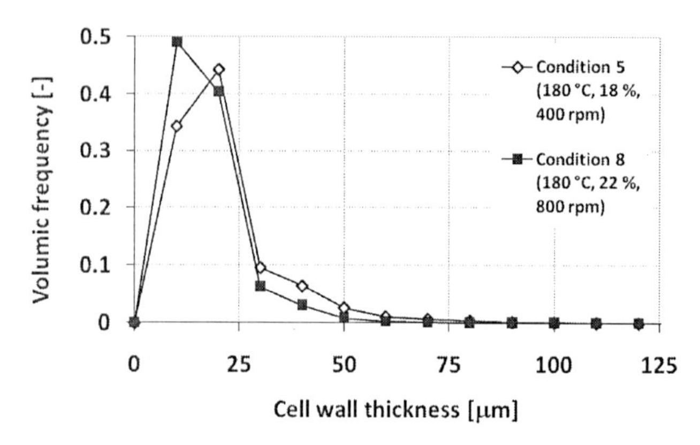

Fig. 7.6: Cell wall thickness distribution of extruded refined flour at condition 5 and at condition 8

Among the two refined flour samples obtained at conditions 5 and at condition 8, the samples extruded at condition 8 showed the highest strength but also the lowest mean cell wall thickness. This is unexpected as the opposite would be expected: for same material properties, larger cell walls would be harder to break. As suggested by Babin et al. (2007), one way to explain this result is to consider the distribution of the cell walls thickness rather than the average cell wall thickness. Comparing the cell wall thickness distribution of the extruded refined flour extruded at condition 5 and at condition 8 (Fig. 7.6), the distribution of the sample extrudes at condition 5 was the most heterogeneous of the two samples. According to the weak link theory (Fasekas et al., 2002) the more heterogeneous the distribution, the higher the probability to encounter a thin wall where the local stress is concentrated and overcome the critical stress, leading to rupture. This may explain the measured weakness of the sample extruded at condition 5 compared to one extruded at condition 8.

As shown in Fig. 7.5 and previously reported in Chapter VI.3.1, the shape of the cells differed according to the process conditions and bran concentration. This may also affect the mechanical properties of the foam. Cell anisotropy was reported to affect the mechanical properties of extruded synthetic polymeric foams (Gibson and Ashby, 1997). In the case of perpendicularly-applied forces, cell stretched in the axial direction would tend to reinforce the structure. On the opposite, cells stretched in the longitudinal direction would weaken the structure (Gibson and Ashby, 1997). The distribution of the cell shape and size within the expanded structures may thus also affect the mechanical properties. As shown in Fig. 7.5, the largest and most spherical cells could be observed at the centre of the extruded shape. Smaller and more anisotropic cells can rather be observed on the sides of the extruded shape. Local stresses may then be different according to the repartition of the cell size and shapes within the extruded sample and may affect the breaking force and pathway. However, it remains difficult to evaluate to which extent the cell shape and distribution within the structure affected the mechanical properties of these samples.

VII.3.3. Effect of cellular structure on mechanical properties of foams at different bran concentration

A similar approach as applied to samples with the same composition was used to investigate the effect of bran on the cellular structure and its relationship with the mechanical properties at same relative density. At similar relative density, increasing the bran concentration to the low level led to a higher stress at rupture and a higher elastic modulus (Fig. 7.4). Observation was made when comparing the refined flour sample extruded at condition 1 (120 °C, 18 % water content in the feed and 400 rpm) with the low bran sample extruded at condition 8 (180 °C, 22 % water content in the feed and 800 rpm). Both samples having same relative density (D = 0.065 vs. 0.066) (Appendices, Tab. 10.6). The low bran sample extruded at condition 8 (showed a significant higher stress at rupture than the refined wheat flour sample extruded at condition 1 (σ^* = 1.15 MPa vs. 0.58 MPa). The low bran sample was characterized by a finer structure with a higher cell density (N_c = 2120 cm^{-3} vs. 100 cm^{-3}), a lower mean cell size (MCS = 410 µm vs. 1610 µm) and a lower mean cell wall thickness (MCWT = 25 vs. 70 µm) than the one made of refined flour sample (Appendices, Tab. 10.6). This is therefore consistent with the fact that finer structures would lead to foams with a higher strength, as previously reported within samples of same composition (see Chapter VII.3.1.1).

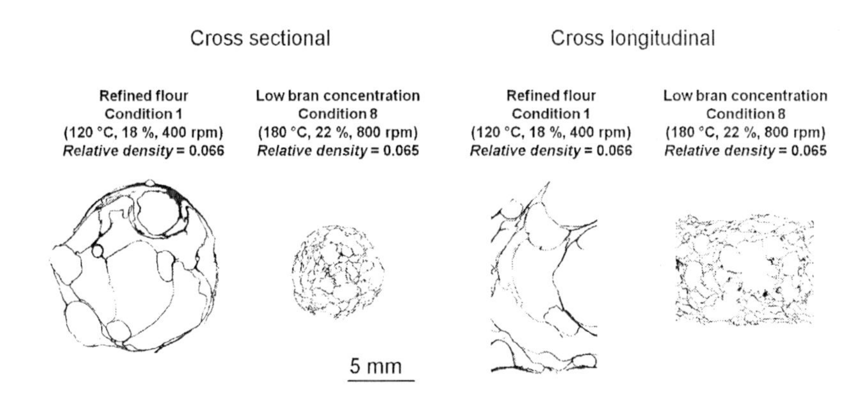

Fig. 7.7: Cross sectional and cross longitudinal X-ray tomography images of extruded samples having a similar relative density (*D*)

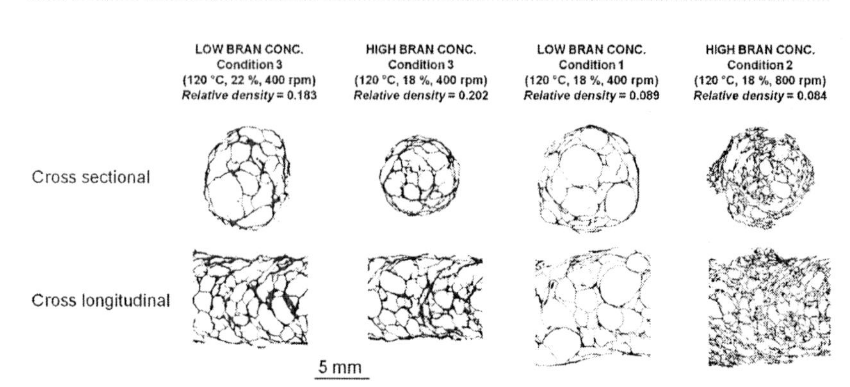

Fig. 7.8: **Cross sectional and cross longitudinal X-ray tomography images of extruded samples at different bran concentration but similar relative density (*D*)**

Further increasing the bran concentration to the high level shifted the power law index of the Gibson-Ashby model towards higher values (Fig. 7.4). The difference in stress at rupture between the low bran and the high bran recipes appeared to depend on the samples relative density. At high relative density, the low bran and high bran samples extruded at condition 3 (LB3 and HB3, 120 °C, 22 % water content and 400 rpm) exhibited close relative density (D = 0.183 and 0.202) (Appendices, Tab. 10.6). The stress at rupture was significantly higher for the high bran sample than for the low bran sample (σ^* = 2.93 MPa vs. 1.80 MPa) (Appendices, Tab. 10.6). The higher stress at rupture measured for the high bran sample (HB3) was associated to a finer structure with a higher density (N_c = 170 cm^{-3} vs. 120 cm^{-3}) of smaller cells (MCS = 580 µm vs. 820 µm) and lower mean cell wall thicknesses (MCWT = 90 µm vs. 110 µm) compared to sample at low bran concentration (LB3) (see Appendices, Tab. 10.6 and Fig. 7.8). On the opposite, at low relative density the stress at rupture of the high bran sample extruded at condition 2 (HB2) was significantly lower than the one of the low bran sample extruded at condition 1 (LB1) (σ^* = 0.55 MPa vs. 0.87 MPa) (Appendices, Tab. 10.6). Both had similar relative density (D = 0.089 vs. 0.084) but the sample with the high bran content (HB2) had a finer structure (N_c = 3650 cm^{-3} vs. 510 cm^{-3} and MCS = 210 µm vs. 1140 µm) than the sample containing low bran (LB1) (see Appendices, Tab. 10.6 and Fig. 7.8). According to the results earlier reported, this finer structure should lead to a higher strength of the extruded body.

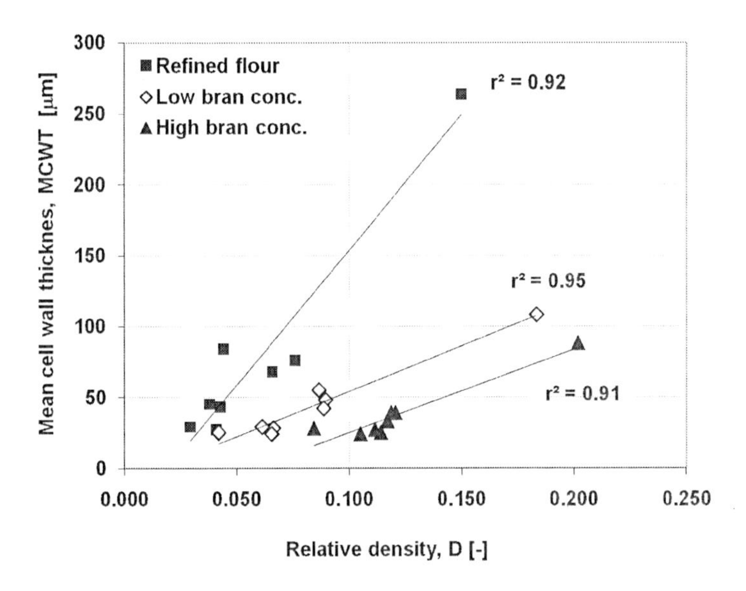

Fig. 7.9: Mean cell wall thickness (MCWT) vs. relative density (*D*) of samples with increasing bran concentration extruded at different conditions

An explanation may be the relationship between the dimensions of the bran particles and those of the cell walls. The cell wall dimensions are depending on the extrusion conditions and bran concentration. The cell wall thickness was affected by the process conditions. It increased linearly with the relative density, for all bran concentrations (correlation factor, $r^2 = 0.92$, 0.95 and 0.91 respectively for refined flour, low bran and high bran recipes) (Fig. 7.9). At constant bran size and aspect ratio of the bran particles, the presence of bran in thinner cell walls is more likely to affect the mechanical properties due to the low adhesion between bran and continuous starch matrix. This hypothesis is illustrated in Fig. 7.10. As previously mentioned the bran particles were rather aligned with the cell walls (see Chapter IV.3.2). In thinner cell walls (Fig. 7.10a) and constant bran particle morphology the ratio of the bran particle thickness to the cell wall thickness would be higher compared to structures with larger thicker cell walls (Fig. 7.10b).

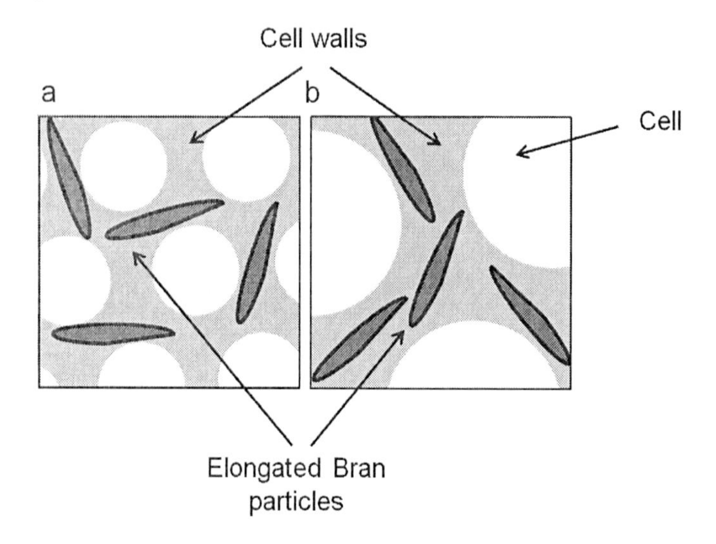

Fig. 7.10: **Schematic illustration of the distribution of elongated bran particles in extruded wheat flour matrices having a similar relative density and highlighting the position of bran particles in thick walled cells (a) or thin walled cells (b)**

These observations allow explaining the changes in mechanical behavior between the low bran and high bran concentration samples depending on their relative density. The low bran sample extruded at condition 1 (LB1) and the high bran sample extruded at condition 2 (HB2) have low relative densities ($D \sim 0.065$). They also have and low mean cell wall thicknesses (MCWT = 30 µm and 25 µm, respectively). The low and high bran samples extruded at condition 3 (LB3 and HB3) had a higher relative density ($D \sim 0.190$). They were characterized by a higher mean cell wall thickness (MCWT = 110 µm and 90 µm, respectively). For the sample LB1 and HB2, showing low mean cell wall thickness, the ratio between the bran particle thickness and the mean cell wall thickness is higher than the one for the samples LB3 and HB3. Thus, the effect of the bran on the mechanical properties of the cell walls may be more significant for samples LB1 and HB2 than for samples LB3 and HB3. As the bran fraction in the sample HB2 sample is higher than in sample LB1, the cell walls are weaker. Thus the sample HB2 exhibits a lower stress at rupture, despite its finer cellular structure. The samples LB3 and HB3 were characterized by higher cell wall thicknesses. Thus the effect of the bran particle on the cell walls was reduced. Therefore, the finer structure obtained when increasing the bran concentration was responsible for the higher stress at rupture obtained for sample HB3 compared to sample LB3.

VII.4. Conclusions

Regardless of the bran concentration, both the stress at rupture and the elastic modulus of the extruded solid foams are positively correlated with their samples relative densities. According to Gibson-Ashby model, the samples without bran and at low bran content have mechanical properties closer to those of structures including open-cells. The high bran samples show mechanical properties closer to those of closed-cell structures. At similar relative densities and sample composition, finer structures with a higher density of small cells lead to a higher mechanical strength.

Increasing the bran concentration significantly decreases the stress at rupture and the elastic modulus of the unexpanded material only at the highest bran concentration. When increasing the bran to an intermediate concentration, the strength of the extruded foams is increased. This is attributed to the finer structures obtained at low bran concentration. The effect of further increasing the bran concentration is dependent on the relative density the foam and the ratio between the bran particle and the cell wall thickness. At higher relative densities, the strength of the structure is further increased due to the even finer structures generated when increasing the bran concentration. At lower relative densities, although finer structures are also obtained, the strength of the structure is reduced compared to the ones at the intermediate bran concentration. This can be caused by the reduced stability of the material induced by the lower cell wall thickness at low relative density and low adhesion between bran and starch.

CHAPTER VIII. CONCLUSIONS AND WAY FORWARD

Consumers demand for healthier food products remains high. Wheat bran, a readily available and low cost by-product, contains a high amount of dietary fibers reported to deliver nutritional and health benefits. Nevertheless, its valorization in extruded food products is limited due to its small expansion volumes and negative effect on textural properties. In order to enhance the use of wheat bran, it is important to understand and control its effect on the parameters driving textural properties of extruded products.

The objective of this work was to investigate the effect of bran on the parameters driving the formation of extruded starchy foams and on their mechanical properties. For this purpose, refined wheat flour was enriched with wheat bran. The recipes were extruded at varying conditions of screw speed, barrel temperature and water content in the feed according to an experimental plan. The effect of bran on the drivers of expansion was assessed using micro-computed X-ray tomography, on-line melt rheology and dynamical scanning calorimetry. The glass transition temperatures and physicochemical properties of the starch were determined and the mechanical properties were assessed using three-point bending tests.

Following the findings of this work and based on the scientific gaps and hypothesis reported in Chapter II, the following conclusions can be made for the range of applied conditions:

Incorporation of bran significantly reduces expansion volumes and leads to finer cellular structures: At same process conditions, increasing the wheat bran concentration in extruded wheat flour significantly increases the longitudinal expansion at the expanse of the sectional and volumetric expansion. The highest sectional expansion of the extruded refined flour samples cannot be reached when adding bran and varying the conditions of extrusion. A relationship is shown between the bulk expansion properties (sectional, longitudinal and volumetric expansion indices) and the cellular structure, regardless of the bran concentration. As the bulk expansion properties can be easily measured, this enables to qualitatively estimate the cellular structure of the samples which requires a more complicating and time-consuming tool. At same volumetric expansion, a favored sectional expansion is associated with coarse cellular structures and cell shapes close to oblate ellipsoids. On the opposite, samples with a bulk expansion closer to isotropy are associated with fine cellular structures and shapes closer to spheres. At same volumetric expansion, increasing the bran concentration favors longitudinal bulk expansion and finer cellular structures.

Increasing bran concentration significantly increases the mechanical strength of extruded foams: At unchanged extrusion conditions, increasing the bran concentration significantly decreases the force necessary to break the extruded samples. The correlation between the diameter of the samples and the breaking force is not satisfying. This indicates that other parameters than the product dimension governs the mechanical properties of the samples. Among them, the stress at rupture and elastic modulus of the foams are positively correlated with the relative density following the model of Gibson-Ashby for solid foams. Nevertheless, the fit with the model is not perfect and samples with similar relative densities exhibit different stresses at ruptures. At same relative density, finer structures are harder to break than coarse structures. The stress at rupture of the cell wall material, obtained by thermomolding of the extruded samples, is significantly decreased only at the highest bran concentration. Nevertheless, at same relative density, increasing the bran concentration increases the stress at rupture. This is attributed to the finer structures generated by addition of bran. Further increasing the bran concentration appears to change the cell connectivity of the structures. According to the Gibson-Ashby model, the refined flour and low bran concentration samples exhibit mechanical properties closer to those of open-cell structures. At the highest concentration of bran, the mechanical properties are closer to those of closed-cell structures. The texture of extruded products based on refined wheat flour is preferred by consumers compared to the texture of extruded products containing wheat bran (e.g. whole wheat flour). This may be linked to the harder structures, composed of a high density of small cells, obtained when adding wheat bran to extruded refined wheat flour. Therefore in order to obtain bran-containing extruded products with textural properties closer to those of extruded products made of refined flour, generation of extruded structures with low density of cells and large cells should be targeted.

Increasing bran significantly affects the degree of nucleation, resulting in finer cellular structures: In order to explain the lower volumetric and sectional expansion volumes obtained with the increasing bran concentration, the effect of bran on each steps of the expansion mechanism was investigated. The density of cells in the final samples is increased when increasing the bran concentration. This can be related to a higher bubble nucleation at the die exit. The higher degree of nucleation can be explained by the presence of non wetted insoluble bran particles in the continuous starch phase. The presence of bran is likely to favor heterogeneous nucleation at the die exit. This may modify the melt rheology due to the formation of thinner bubble walls. These thinner bubble walls can induce less resistance against growth. The higher density of growing bubbles in the melt may also increase the shear at the interface between the bubbles, reducing the melt viscosity.

Addition of bran modifies the melt shear viscosity, which may explain the reduced volumetric and sectional expansion: The growth of bubbles is reduced when increasing the bran concentration. This is shown by the lower maximum sectional expansion index at the die,. The growth of bubbles at the die exit is driven by the vaporization of water and by the shear and extensional forces of the melt. Indeed, to grow the bubble has to overcome the resistance created by the shear viscosity. In the tested range of shear rates, the shear viscosity of the melt is significantly increased only at the highest bran concentration. This indicates a threshold value from which the shear viscosity starts to be affected by the bran concentration. At high bran concentration, the decrease in free volume of the starch molecules induced by the presence of bran particles can explain this effect. Physical interactions between the bran particles may also participate to the increased shear viscosity of the matrix. This increase in shear viscosity may hinder the growth of the bubble. Addition of bran significantly modifies the entrance pressure drop, measured using the orifice die method (Bagley plot). This may indicate reduced elastic properties of the melt. This reduction in elastic properties may reduce extensibility of the membrane of the bubbles when they undergo biaxial extension during growth. Nevertheless, further work is necessary to conclude on the effect of wheat bran on the extensional viscosity of starchy melts. At same water content or water activity, the glass transition temperature of the starch phase is reduced when increasing the bran concentration. This is due to higher free water available for the starch phase, resulting from the lower hygrocapacity of bran fibers. This decrease in glass transition temperature may induce a lower viscosity of the starch phase. This may counteract the increase in shear viscosity of the matrix attributed to the bran particles.

Interactions between starch and bran particles can affect the melt rheology and expansion properties: Physicochemical properties of starch drive the rheological properties of the melt. After extrusion, the crystalline structure of starch is lost, irrespective of the process conditions and bran concentration. Although the crystallinity of starch is lost, an amorphous structure that could still hydrate and burst under shear during rapid visco analysis remains after extrusion. The amount of this amorphous structure depends on the process conditions (decreasing with the specific mechanical energy) and the macromolecular state of amylopectin. Strong interactions between the bran concentration and the process conditions exist and can influence the melt rheological properties. Several factors may explain these interactions. The solubility of starch in cold water was estimated after extrusion from the water solubility index. This value only increases at the highest bran concentration. This means an increase in starch transformation at this level of bran. As a higher degree of starch transformation was shown to reduce the melt shear viscosity, this may also counteract the increase in shear viscosity of the bran-containing matrix associated to the presence of bran particles. Bran is not fully inert. It was shown to exhibit an elastoplastic behavior. Its mechanical properties may therefore be affected by the temperature and moisture content. Additionally, the bran particle size is reduced when increasing the shear stress in the extruder.

Bubble burst occurs during growth and can significantly reduce expansion: The surface porosity of the extruded samples is increasing with an increasing concentration of bran. This higher surface porosity is reflecting an increased burst of bubbles at the sample surface. As the pressure is increasing in the bubble, its membrane cannot withstand the pressure anymore due to rupture at the interface between the continuous starch matrix and the bran particles. The rupture is likely favored by a low bubble membrane thickness with a thickness close to the size of the bran particles. As high volumetric expansions and high bran content favors lower wall thickness, these conditions may favor rupture of the bubbles. Additionally, the bran particle is varied according to the process conditions. Large bran particles remaining after extrusion may also favor rupture of the bubbles walls during bubble growth.

Shrinkage during expansion occurs and can contribute to a reduction in expansion volumes: Regardless of the bran concentration, the samples exhibited shrinkage after maximum sectional expansion at the die. At same process conditions, the maximum sectional expansion index of the bran-containing samples was lower than the sectional expansion index (after shrinkage) of the refined flour. Therefore a higher shrinkage of the structure of the bran-containing samples could not explain the reduced final expansion properties of these samples compared to those of the extruded refined flour. Shrinkage occurs when the bubble membrane cannot withstand the pressure inside the bubble anymore. It is favored at low melt shear viscosity. It is also likely than thinner bubble walls also favor shrinkage. The elastic properties may also be involved in the shrinkage phenomena due to elastic stress relaxation effects.

This work opens questions for possible future research on bran-containing extruded products: A combination of interdependent parameters was shown to explain the reduced expansion volumes obtained when incorporating wheat bran in extruded products. It is therefore, difficult to quantify the individual effect of each parameter. These parameters are driven by the starch and bran properties and their interactions depending on the extrusion conditions. The melt rheological properties were assessed using on-line rheology. Nevertheless, the range of apparent shear rates ($5 - 30\ s^{-1}$) obtained with the rheometer was much lower than the value of shear rate at the die (about $600\ s^{-1}$). Depending on the melt flow behavior this may result in significant differences in shear viscosity at higher shear rates. Additionally, the effect of bran particle may be different at higher shear rate. Improvement in the design of the twin-slit rheometer should enable to reach the shear rates achieved in industrial die designs. Single capillary/slits rheometers allow covering a wider range of apparent shear rates. They may also be used. Nevertheless, they do not allow controlling the ingredient themomechanical history in the rheometer.

The elastic properties of the melt appear to play a major role in the expansion mechanism. The use of off-line techniques to measure elastic properties such as lubricated squeezing flow rheometry or dynamic mechanical thermal analysis (DMTA) is challenging when using conditions of water and temperature encountered during extrusion. The main challenges are caused by water evaporation and very high viscosity. Lubricated squeezing flow rheometry also requires cohesive dough for the measurement. This would therefore not be applicable for the high bran concentration samples. The use of on-line techniques remains the most appropriate way to measure the melt elastic properties in the conditions of extrusion. The entrance pressure drop obtained using the Bagley plot is a qualitative measurement of the extensional viscosity. Complex calculations using several assumptions would be necessary to extract the extensional viscosity from the Bagley plot. Further work should be carried to characterize and fully understand the effect of wheat bran on the elastic properties of extruded starchy melts using as an inspiration, the work performed in the plastic/fillers extrusion.

As reported in literature, wheat bran exhibits an elastoplastic behavior. The mechanical properties of the bran particles may be changed by the process conditions. Further work should investigate the mechanical/rheological properties of pure wheat bran. This would allow characterizing its rheological behavior during extrusion of bran-containing products. This may also allow finding solutions to modify bran prior or during extrusion and improve its expansion properties. Additionally, its particle size was modified during extrusion depending on the process conditions. Therefore the effect of bran particle size on melt rheology and expansion properties not only depends on the particle size of the starting material but also on the degree of size reduction during extrusion. The use of synthetic fillers as model particles may enable to avoid particle size changes during extrusion. Thus enable to investigate the effect of bran particle size on expansion.

This work may contribute to finding solutions to improve the expansion properties of extruded bran-containing products: Reducing the particle size of fibers has been proposed as a solution to increase the expansion of extruded products containing dietary fibers (see Chapter II.5). Such effect was attributed to the reduction in the degree of bubble bursting at low fiber particle size and to the higher water-binding capacity of low fiber particle sizes. As shown in this study, rupture of bubble membranes is favored at a critical thickness close to the size of bran particle. In the case of the extruded samples based on the pure refined flour, the mean cell wall thickness ranged from 30 µm to 75 µm. Therefore, the bran particles should be within this range to avoid rupture of the bubble membrane during growth. To obtain such low particle size, traditional grinding is not enough. More sophisticated and expensive techniques are necessary. Additionally, decrease in particle size was also shown to increase the viscosity of starchy dough (Zhang and Moore, 1997). This may hinder the expansion. Another approach to improve the expansion properties of extruded bran-containing starch products would consist in modifying the rheological properties of the bran to obtain closer properties to those of starch. Increasing the compatibility and adhesion properties between the continuous starch phase and the bran properties would also enable to increase the expansion properties of bran-containing matrices. Such approach was proposed by Blake (2006) to improve the expansion properties of corn flour enriched in corn bran. For this the author used an alkali-treatment to hydrolyze the ferulic acid ester cross-linkages in corn bran. The alkali-treatment decreased the shear viscosity and increased the extensional viscosity of the corn flour enriched with treated corn bran, improving expansion. This effect was mainly attributed to an increase in soluble arabinoxylan content.

Wheat bran represents an opportunity for the Food Industry to increase the dietary fiber content of extruded products using a low cost and readily available ingredient. Due to its low price, wheat bran can be used as a source of dietary fibers while maintaining products at reasonable prices. Consumers are aware of the benefits of dietary fibers for their health. They can benefit from an extruded products enriched with wheat bran. Industrial food companies can therefore respond to the consumer demand for healthier foods and increase their sales and profits. Modification of wheat bran using chemical, enzymatic and/or mechanical treatments, complying with the regulation of countries, should in a near future allow improving the textural properties of extruded products containing wheat bran.

REFERENCES

Ablett, S., Attenburrow, G.E. & Lillford, P.J. (1986). The significance of water in the baking process. In J.M.V. Blanshard, P.J. Frazier and T. Galliars (Eds.), *Chemistry and physics of baking: material, process and products*. Royal Society of Chemistry, London, pp. 30-41

Agbisit, R. Alavi, S., Cheng, E., Herald, T. & Trater, A. (2007). Relationships between microstructure and mechanical properties of cellular cornstarch extrudates. *Journal of Texture Studies*, 38, 199-219

Alaoui, A.H., Woignier, T., Scherer, G.W. & Phalippou, J. (2008). Comparison btween flexural and uniaxial compression tests to measure the elastic modulus of silica aerogel. *Journal of Non-Crytalline Solids*, 354, 4556-4561

Antoine, C., Peyron, S., Mabille, F., Lapierre, C., Bouchet, B., Abecassis, J. & Rouau, X. (2003). Individual Contribution of Grain Outer Layers and Their Cell Wall Structure to the Mechanical Properties of Wheat Bran. *Journal of Agricultural and Food Chemistry*, 51, 2026-2033

Anonymous (2009). *Codex: Guidelines for use of nutrition and health claims.*

Alvarez-Martinez, L., Kondury, K.P. & Harper, J.M. (1988). A general model for expansion of extruded products. *Journal of Food Science,* 53 (2), 609–615

Anderson, R.A., Conway, H.F., Pfeifer, V.F. & Griffin, E.L. (1969). Gelatinization of corngrits by roll- and extrusion-cooking. *Cereal Science Today,* 14 (1), 4–11

Armitage, P., & Colton, T. (1998). *Encyclopedia of Biostatistics* (vol. 2, pp. 1431). Wiley Editions, UK

Babin, P., Della Valle, G., Dendeviel, R., Lourdin, R. & Salvo, L. (2007). X-ray tomography study of the cellular structure of extruded starches and its relations with expansion phenomenon and foam mechanical properties. *Carbohydrate Polymers*, 68, 329-340

Bagley, E.B. (1957). End corrections in the capillary flow of polyethylene. *Journal of Applied Physics*. 28, 193-209

Barrès C., Vergnes B., Tayeb J. & Della Valle G. (1990). Transformation of wheat flour by extrusion-cooking: influence of screw configuration and operating conditions. *Cereal Chemistry*, 67, 427-433.

Barrett, A.H., Rosenberg, S. & Ross, E.W. (1994). Fracture intensity distributions during compression of puffed corn meal extrudates: method for quantifying fracturability. *Journal of Food Science*, 59, 617-620

Barron, C., Surget, A. & Rouau, X. (2007). Relative amount of tissues in mature wheat (*Triticum aestivum, L.*) grain and their carbohydrate and ohenolic acid composition. *Journal of Cereal Science*, 45, 88-96

Bhattacharya, M., Padmanabhanm, M. & Seethamraju, K. (1994). Uniaxial extensional viscosity during extrusion cooking from entrance pressure drop method. *Journal of Food Science*, 59 (1), 221-226

Bhattacharya, S., Sudha, M.L. & Rahim, A. (1999). Pasting characteristics of an extruded blend of potato and wheat flours. *Journal of Food Engineering*, 40, 107-111

Binding, D.M. (1988). An approximate analysis for contraction and converging flows. *Journal of Non-Newtonian Fluids Mechanics*, 27, 173-189

Björck, I., Nyman, M. & Asp, N-G. (1984). Extrusion cooking and dietary fiber: effects on dietary fiber content and degradation in the rat intestinal tract. *Cereal chemistry*, 61 (2), 174-179

Blake, O. (2006). Effect of molecular and supramolecular characteristics of selected dietary fibers on extrusion expansion. *PhD Dissertation*, Purdue University, West Lafayette, Indiana

Bouzaza, D., Arhaliass, A. & Bouvier, J. M. (1996). Die design and dough expansion in low moisture extrusion-cooking process. *Journal of Food Engineering*, 29, 139-152

Brasseur, E., Fyrillas, M.M., Georgiou, G.C. & Crochet, M.J. (1998). The time-dependant extrudate-swell problem of an Oldroyd-B fluid with slip along the wall. *Journal of Rheology*, 42 (3), 549-566

Breen, M.D., Seyam, A.A. & Banasik, O.J. (1977). The effect of mill by-products and soy protein on the physical characteristics of expanded snack foods. *Cereal Chemistry*, 54 (4), 728-736

Brennan, M., Monro, J. A, & Brennan, C. S. (2008). Effect of inclusion of soluble and insoluble fibres into extruded breakfast cereal products made with reverse screw configuration. *International Journal of Food Science and Technology*, 43, 2278-2288

Brümmer, T., Meuser, F., van Lengerich, B. & Niemann, C. (2002). Effect of extrusion cooking on molecular parameters of corn starch. *Starch/Stärke*, 54(1), 9-15

Burt, D.J., & Russel, P.L. (1983). Gelatinization of low water content wheat starch-water mixtures. *Starch/Stärke*, 35, 354-360

Cervone, N.H. & Harper, J.M. (1978). Viscosity of an intermediate moisture dough. *Journal of Food Processing Engineering*, 2, 83-95.

Chan, Y., White, J.L. & Oyanagi, Y. (1978). Influence of glass fibers on the extrusion and injection molding characteristics of polyethylene and polystyrene melts. *Polymer Engineering and Science*, 18 (4), 268-272

Chanvrier, H., Della Valle, G., & Lourdin, D. (2006). Mechanical behaviour of corn flour and starch-flour: A matrix-particle interpretation. *Carbohydrate Polymers*, 65, 346-356

Chang, Y.P., Cheah, P.B. & Seow, C.C. (2008). Plasticizing-Antiplasticizing effects of water on physical properties of tapioca starch films in the glassy state. *Food Engineering and Physical Properties*, 65 (3), 445-451

Champenois, Y., Colonna, P., Buléon, A., Della Valle, G. & Renault, A. (1995). Gélatinisation et rétrogradation de l'amidon dans le pain de mie. *Science des Aliments*, 15, 593-614

Chinnaswamy, R. (1993). Basics of cereal starch expansion. *Carbohydrate Polymers*, 21, 157-167

Cho, S.S. & Clark, C. (2001). *Wheat bran: physiological effects*. In Cho, S.S., Dreher, M.L (Eds.). Handbook of Dietary Fiber. Marcel Dekker: New York

Christianson, D.D., Hodge, J.E., Osborne, D. & Detroy, R.W. (1981). Gelatinization of wheat starch as modified by xanthan gum, guar gum and cellulose gum. *Cereal Chemistry*, 58 (6), 513-517

Chungcharoen, A. & Lund, D.B. (1987). Influence of solutes and water on rice starch gelatinization. *Cereal Chemistry*, 64 (4), 240-243

Cisneros, F.H. & Kokini, J.L. (2002a). A generalized theory linking barrel fill length and air bubble entrapment during extrusion of starch. *Journal of Food Engineering*, 51 (2), 139-142

Cisneros, F.H. & Kokini, J.L. (2002b). Effect of extrusion operating parameters on the air bubble entrapment. *Journal of Food Engineering*, 25 (4), 251-283

Cogswell, F.N. (1972). Converging flow of polymer melts in extrusion dies. *Polymer Engineering Science*, 12, 64-73

Colonna, P., Doublier, J. L ., Melcion, J. P., de Monredon, F. & Mercier, C. (1984). Extrusion cooking and drum drying of wheat starch. I. Physical and macromolecular modifications. *Cereal Chemistry*, 61 (3), 538-543

Cornell, H. & Hoveling, A.W. (1998). *Wheat: chemistry and utilization*. Technomic Publishing Company, Inc., Lancaster, Pennsylvenia, pp. 426

Crowson, R.J., Folkes, M.J., & Bright, P.F. (1980). Rheology of short glass fiber-reinforced thermoplastics and its application to injection molding. I. Fiber motion and viscosity measurement. *Polymer Engineering and Science*, 20 (14), 925-933

Cuq, B., Abecassis, J. & Guilbert, S. (2003). State diagrams to help describe wheat bread processing. *International Journal of Food Science and Technology*, 38, 759-766

Davidson, V. J. Paton, D., Diosady, L. L. & Larocque, G. (1984). Degradation of wheat starch in a single screw extruder: characteristics of extruded starch polymers. *Journal of Food Science*, 49, 453-458

Delcour, J.A., & Hoseney, R.C. (2010). *Principles of cereal science and technology*. 3rd Ed. AACC International: St Paul, Minesota, pp. 270

Della Valle, G., Colonna, P. & Patria. (1996). Influence of amylose content on the viscous behaviour of low hydrated molten starches. *Journal of Rheology*, 40 (3), 347-362

Della Valle, G., Vergnes, B., Colonna, P. & Patria, A. (1997). Relations between rheological properties of molten starches and their expansion behavior in extrusion. *Journal of Food Engineering*, 31, 277–296.

Desrumaux, A., Bouvier, J.M. & Burri, J. (1998). Corn grits particle size and distribution: effects on the characteristics of expanded extrudates. *Journal of Food Science* 63 (5), 1–7.

Dhanaskharan, M. & Kokini, J.L. (2000). Viscoelastic flow modeling in the extrusion of a dough-like fluid. *Journal of Food Process Engineering*, 23, 237-247

Ding, Q.-B., Ainsworth, P., Tucker, G. & Marson, H. (2005). The effect of extrusion conditions on the physicochemical properties and sensory characteristics of rice-based expanded snacks. *Journal of Food Engineering*, 66, 283-289

Donovan J.W. (1979). Phase transitions of the starch-water systems. *Biopolymers*, 18, 263-275

Dreher, M.L. (2001). *Dietary fiber overview.* In Cho, S.S., Dreher, M.L. (Eds.). Handbook of Dieter Fiber. Marcel Dekker Inc.: New York, pp. 868

Dubois, M., Gilles, K. A., Hamilton, J. K., Rebers, P. A. & Smith, F. (1956). Colorimetric method for determination of sugars and related substances. *Analytical Chemistry*, 28, 350

Eastman, J., Orthoefer, F. & Solorio, S. (2001). Using extrusion to create breakfast cereal products. *Cereal Food World*, 46 (10), 468-471

Falcone, R.G. & Phillips, R.D. (1988). Effects of feed composition, feed moisture, and barrel temperature on the physical and rheological properties of snack-like products prepared from cowpea and sorghum. *Journal of Food Science*, 53, 1464-1471

Fan, J., Mitchell, J.R. & Blanshard, J.M.V. (1994). A computer simulation of the dynamics of bubble growth and shrinkage during extrudate expansion. *Journal of Food Engineering*, 23, 337-356

Food and Agricultural Organization of the United Nations. (2010). FAO cuts wheat production forecast but considers supplies adequate. *FAO Media Center*, www.fao.org

Fasekas, A., Dendeviel, R., Salvo, L. & Brechet, Y. (2002). Effect of microstructural topology upon the stiffness and strength of 2D cellular structures. *International Journal of Mechanical Sciences*, 44, 2047-2066

Garber, B.W., Hsieh, F. & Huff, H.E. (1997). Influence of particle size on the twin-screw extrusion of corn meal. *Cereal Chemistry*, 74 (5), 656-661

Gevaudan, A., Chuzel, G., Didier, S. & Andrieu, J. (1989). Physical properties of cassava mash. *International Journal of Food Science and Technology*, 24, 267-645

Ghiasi, K., Hoseney, R.C. & Varriano-Marston, E. (1983). Effects of flour components and dough ingredients on starch gelatinization. *Cereal Chemistry*, 60 (1), 58-61

Giesbrecht, F.G. & Gumpertz, M.L. (2004). *Planning, Construction and Statistical: Analysis of Comparative Experiments*. Wiley Editions: Hoboken, USA, pp. 693

Gibson, L.J. & Ashby, M.F. (1997). *Cellular Foams*. Cambridge University Press: Cambridge, UK, pp. 510

Gonzàles, R.J., Torres, R.L., de Greef, D.M. & Guadalupe, B.A., 2006. Effects of extrusion conditions and structural characteristics on melt viscosity of starchy materials. *Journal of Food Engineering*, 74, 96-107

Ghosh, P., Debaprasad, D. & Chakrabarti, A. (1997). Reactive melt processing of polyethylene: effect of peroxide action on polymer structure, melt rheology and relaxation behavior. *Polymer*, 38 (25), 6175-6180

Ghosh, P. & Chakrabarti, A. (2000). Effect of incorporation of conducting carbon black as filler on the melt rheology and relaxation behavior of ethylene-propylene-diene monomer (EFDM). *European Polymer Journal*, 36, 607-617

Gordon, M. & Taylor, J. (1993). Ideal Copolymers and the second order transitions of synthetic rubbers: 1. Non crystalline copolymers. *Journal of Applied Chemistry*, 2, 493-500

Greffeuille, V., Mabille, F., Rousset, M., Oury, F.-X., Abecassis, J. & Lullien-Pellerin, V. (2007). Mechanical properties of outer layer from near-isogenic lines of common wheat differing in hardness. *Journal of Cereal Sciences*, 45, 227-235

Gualberto, D. G. Bergman, C. J. Kazemzadeh, M. & Weber, C.W. (1997). Effect of extrusion processing on the soluble and insoluble fiber, and phytic acid contents of cereal brans. *Plant Foods for Human Nutrition*, 51, 187-198

Guy, R.C.F. (1985). The extrusion revolution. *Food manufacture*, 60, 26-29

Guy, R.C.E. & Horne, A.W. (1988). *Extrusion cooking and co-extrusion.* In Blanchard, J.M.V., Mitchell, J.R. (Eds.). Food Structure: its Creation and Evaluation. Butterworth: London, pp. 331-349

Guy, R. (2001). *Extrusion cooking: technologies and applications*. CRC Press: New York, Washington DC, pp. 206

Hagenimana, A., Ding, X. & Fang, T. (2006). Evaluation of rice flour modified by extrusion cooking. *Journal of Cereal Science*, 43, 38-46

Hamaker, B. R. (2008). *Technology of functional cereal products*. CRC Press, New York, pp. 548

Han, C. D. (1974). On slit- and capillary-die rheometry. *Journal of Rheology*, 18, 163–190

Harper, J.M., Rhodes, T.P. & Wanninger, L.A. (1971). Viscosity model for cooked cereal dough. *AIChe Symposium Series*, No 108

Hatzikiriakos, S.G. & Mitsoulis, E. (1996). Excess pressure losses on the capillary flow of molten polymers. *Rheologica Acta*, 35, 545-555

Hemery, Y.M., Mabille, F., Martelli, M.R. & Rouau, X. (2010). Influence of water and negative temperatures on the mechancial properties of wheat bran and its constitutive layers. *Journal of Food Engineering*, 98, 360-369

Hildebrand, T. & Rüegsegger, P. (1997). A new method for the model-independent assessment of thickness in three-dimensional images. *Journal of Microscopy*, 187 (1), 67-75

Hohenberger, W. (2009). *Fillers and reinforcement / coupling agent.* In H. Zweifel, R.D. Maier, M. Schiller (Eds.). Plastics additives handbook (6th Ed). Carl Hanser Verlag.: Munich, pp. 919-1028

Horvat, M. Hirth, M., Emin, A., Schuchmann H. P., Hochstein B. & Willenbacher, N. (2009). Online-Rheologie zur Produktentwicklung extrudierter, funktioneller Zerealien. *Chemie Ingenieur Technik*, 81 (8), 1173-1180

Hoseney, R.C., Mason, W.R, Lai, C.S. & Guetzlaff, J. (1992). Factors affecting the viscosity and structure of extrusion-cooked wheat starch. In Kokini, J.L., Ho, C.T., Karwe, M.V., *Food extrusion science and technology*. Marcel Dekker. Inc.: New York, p. 277-305

Hsieh, F., Mulvaney, S.J., Huff, Lue, S. & Brent, J. (1989). Effect of dietary fiber and screw speed on some extrusion processing and products variables. *Lebensmittel, Wissenschaft und Technologie*, 22, 204-207

Hsieh, F., Huff, H.E.E. & Stringer, L. (1991). Twin-screw extrusion of sugar beet fiber and corn meal. *Lebensmittel, Wissenschaft und Technologie*, 24, 495-500

Jang, J.K. & Pyun, Y.R. (1996). Effect of moisture content on the melting of wheat starch. *Starch/Stärke*, 48 (2), S48-51

Kaletunç, G. & Breslauer, K.J. (1996). Construction of a wheat-flour state diagram: application to extrusion processing. *Journal of Thermal Analysis*, 47, 1267-1288

Kalichevsky, M.T., Jaroszkiewicz, E.M., Ablett, S., Blanshard, J.M.V. & Lillford, P.J. (1992). The glass transition of amylopectin measured by DSC, DMTA and NMR. *Carbohydrate Polymers*, 18, 77-88

Kaur, K., Singh, K., Sekhon, K. & Singh, B. (1999). Effect of hydrocolloids and process variables on the extrusion behaviour of rice grits. *Journal of Food Science and Technology*, 36, 127-132

Klingler, R.W., Meuser F. & Niediek, E.A. (1986). Effect of the form of energy transfer on the structural and functional characteristics of starch. *Starch/Stärke*, 38, 40-44

Kokini, J.L., Chang, C.N. & Lai, L.S. (1992). The role of rheological properties on extrudate expansion. In Kokini, J.L., Ho, C.-T. and Karwe, M.V. editors. *Food extrusion science and technology*. New York, N.Y.: Marcel Dekker Inc. p 631-653

Kokini, J.L., Cocero, A.M., Madeka, H. & de Graaf, E. (1994). The development of state diagrams for cereal proteins. *Trends in Food Science & Technology*, 5, 281-288

Krieger, I.M. & Dougherty T.J. (1959). A mechanism for non-Newtonian flow in suspensions of rigid spheres. *Transactions of the Society of Rheology*, 3, 137–152.

Lach, L. (2006). *Modeling vapor expansion of extruded cereals*. PhD dissertation, Lausanne, Switzerland: EPFL 250 p. Available from EPFL n°3476

Lee, S. T. (2000). *Foam nucleation in gas-dispersed polymeric systems*. In S.T. Lee (Ed.), *Foam Extrusion*. Technomic Pub. Co.: Lancaster, pp.81-144

Lei, M., Lee & T.–C. (1996). Effect of extrusion temperature on solubility and molecular weight disruption of wheat flour proteins. *Journal of Agriculture Food Chemistry*, 44, 763-768

Leung, S.N.S. (2009). Mechanisms of cell nucleation, growth and coarsening in plastic foaming: theory, simulation, and experiments. *PhD dissertation*, University of Toronto

Lobe, V.M. & White, J.L. (1979). An experimental study on the influence of carbon black on the rheological properties of a polystyrene melt. *Polymer Engineering & Science*, 19 (9), 617-624

Lou, J. & Harinath, V. (2004). Effects of mineral fillers on polystyrene melt processing. *Journal of Materials Processing Technology*, 152, 185-189

Lourdin, D., Della Valle, G. & Colonna, C. (1995). Influence of amylose content on starch films and foams. *Carbohydrate Polymers*, 27, 251-270

Lue, S., Hsieh, F., Peng, I.C. & Huff, H.E. (1990). Expansion of corn extrudates containing dietary fiber: A microstructure study. *Lebensmittel Wissenschaft und Technology*, 23, 165-170

Lue, S., Hsieh, F. & Huff, H.E. (1991). Extrusion cooking of corn meal and sugar beet fiber: effects on expansion properties, starch gelatinization and dietary fiber content. *Cereal Chemistry*, 68 (3), 227-234

Macosko, C.W. (1994) *Rheology: principles, measurements, and applications*. Wiley-VCH, New York

Marlett, M., McBurney, M. I. & Slavin, J. L. (2002). Position of the American Dietetic Association: health implications of dietary fiber. *Journal of the American Dietetic Association*, 102 (7), 993-1000

Martin, O., Averous, L. & Della Valle, G. (2003). In-line determination of plasticized wheat starch viscoelastic behaviour: impact of processing. *Carbohydrate Polymers*, 53, 169-182

Minagawa, N. & White, J.L. (1976). The influence of titanium dioxide on the rheological and extrusion properties of polymer melts. *Journal of Applied Polymer Science*, 20, 501-523

Mitsoulis, E. & Hatzikiriakos, S.G. (2003). Bagley correction: the effect of contraction angle and its prediction. *Rheologica Acta*, 42, 309-320

Mooney, M. (1931). Explicit Formulas for Slip and Fluidity. *Journal of Rheology*, 2 (2), 210-216

Moore, D., Sanei, A., van Hecke E. & Bouvier, J.M. (1990). Effect of Ingredients on Physical/Structural Properties of Extrudates. *Journal of Food Science*, 55 (5), 1383-1387

Moraru, C. I. & Kokini, J. L. (2003). Nucleation and expansion during extrusion and microwave heating of cereal foods. *Comprehensive reviews in food science and food safety*, 2, 147-165

Morrison, W.R., Tester, R.F., Snape, C.E., Law, R. & Gidley, M.J. (1993). Swelling and gelatinization of cereal starches: IV. Some effects of lipid-complexed amylose and free amylose in waxy and normal barley starches. *Cereal Chemistry*, 70 (4), 385-391

Núñez, M, Sandoval, A.J., Müller, A. J., Della Valle, G. & Lourdin, D. (2009). Thermal characterization and phase behavior of a ready-to-eat breakfast formulation and its starchy components. *Food Biophysics*, 4, 291-303

Ofoli, R.Y. & Steffi, J.F. (1993). Some observations on the use of slit rheometry for characterizating the primary normal stress difference of extrudates. *Jounral of Food Engineering*, 18, 145-157

Onwulata, C.I., Konstance, R.P., Smith, P.W. & Holsinger, V.H. (2001). Co-extrusion of dietary fiber and milk proteins in expanded corn products. *Lebensmittel, Wissenschaft und Technologie*, 34, 424-429

Orford, P. D., Parker, R. & Ring, S. G. (1990). Aspects of the glass transition behavior of mixtures of carbohydrates of low molecular weight. *Carbohydrate Research*, 196, 11-18

Osswald, T.A. & Menges, G. (1995). *Materials Science of Polymers for Engineers* (2nd Ed.). Hanser Publishers, Munich, pp. 622

Padmanabhan, M. & Bhattacharya, M. (1991). Flow behavior and exit pressures of corn meal under high-shear-high-temperature extrusion conditions using a slit die. *Journal of Rheology*, 35, 315-342

Padmanabahan, M & Bhattacharya, M. (1993). Planar extensional visocisty of corn meal dough. *Journal of Food Engineering*, 18, 389-411

Pai, D. A., Blake, O. A., Hamaker, B. R. & Campanella, O. H. (2009). Importance of extensional rheological properties on fiber-enriched corn extrudates. *Journal of Cereal Science*, 50, 227-234

Palzer, S. (2004). Bedeutung und Berechnung des Glasübergangs komplexer amorpher Lebensmittelkomponenten Teil II. *Lebensmittletechnic*, 9, 70

Parker, R., Ollett, A.-L., Lai-Fook, R. A. & Smith, A. C. (1989). *The rheology of food melts and its application to extrusion processing*. In R. E. Carter (Ed.), Rheololy of Food, Pharmaceutical and Biological Materials. Elsevier: London, pp. 57-73.

Peyron, S., Chaurand, M., Rouau, X. & Abecassis, J. (2002). Relationship between bran mechanical properties and milling behavior of durum wheat. Influence of tissue thickness and cell wall thickness. *Journal of Cereal Science*, 36 (3), 377-386

Phillips, G. & Cui, S.W. (2011). An introduction: Evolution and finalization of the regulatory definition. *Food Hydrocolloids*, 25, 139-143

Pomeranz, Y. (1988). *Wheat Chemistry and Technology* (3rd Ed.). AACC, Inc., St Paul, Minnesota, pp. 426

Politz, M. L. Timpa, J. D., White, A. R. & Wasserman, B. P. (1994). Non-aqueous gel permeation chromatography of wheat starch in deimethylacetamide (DMAC) and LiCl: extrusion-induced fragmentation. *Carbohydrate polymers*, 24, 91-99

Krieger, I.M. & Dougherty, T.J. (1959). A mechanism for non-Newtonian flow in suspensions of rigid spheres. *Transaction of the Society of Rheology*, 3, 137–52.

Rallet, M.-C. Thibault, J.-F. & Della Valle, G. (1990). Influence of extrusion-cooking on the physic-chemical properties of wheat bran. *Journal of Cereal Science*, 11, 249-259

Robin, F., Engmann, J., Pineau, N., Chanvrier, H., Bovet, N. & Della Valle, G. (2010a). Extrusion, structure and mechanical properties of complex starchy foams. *Journal of Food Engineering*, 98, 19-27

Robin, F., Engmann, J., Tomasi, D., Breton, O., Parker, R., Schuchmann, H. P., Palzer, S. (2010b). Adjustable twin-slit rheometer for shear viscosity measurement of extruded complex starchy melts, *Chemical Engineering & Technology*, 33 (10), 1672-1678

Rhodes, D.I., Sadek, M. & Stone, B.A. (2002). Hydroxycinnamic acids in walls of wheat aleurone cells. *Journal of Cereal Science*, 36, 67-81

Rothon, R.N. (2003). *Particulate-filled polymer composite* (2nd Ed.). Shawbury: Rapra Technology Ltd

Roudaut, G., Dacremont, C., Vallès Pàmies, Colas, B. & Le Meste, M. (2002). Crispiness : a critical review on sensory and material science approaches. *Trends in Food Science & Technology*, 13, 217-227

Roscoe, R. (1953). *Flow properties of dispersed systems*. Hermans JJ (Ed.). New York: Interscience

Schuchmann, H.P. & Danner, T. (2000). Product Engineering Using the Example of Extruded Instant Powders. *Chemical Engineering and Technology*, 23 (4), 303-308

Senouci, A. & Smith, A.C. (1988a). An experimental study of food melt Rheology. I. Shear viscosity using a slit die viscometer and a capillary rheometer. *Rheologica Acta*, 27, 546-554

Senouci, A. & Smith, A.C. (1988b). An experimental study of food melt rheology. II. End pressure effects. *Rheologica Acta*, 27, 649-655

Shenoy, A.V. (1999). *Rheology of filled polymer systems*. Kluwer Academic Publishers: Dordrecht, pp. 475

Shogren, R.L. (1992). Effect of moisture content on the melting and subsequent physical aging of corn starch. *Carbohydrate Polymers*, 19, 83-90

Smith, A.C. (1992). Studies on the physical structure of starch-based material in the extrusion cooking process. In Kokini, J.L., Ho, C.-T., Karwe, M.V. (Eds.). *Food extrusion science and technology*. Marcel Dekker Inc.: New York, p 573-618

Srichuwong, S. & Jane, J.-L. (2007). Physicochemical properties of starch affected by molecular composition and structures: a review. *Food Science and Biotechnology*, 16 (5), 663-674

Steffe, J. (1996). *Rheological methods in food process engineering*. Freeman Press, East Lansing, Minnesota

Stute, R. (1991). Hydrothermal modifications of starches: the difference between annealing and heat-moisture treatment. *Starch/Stärke*, 6 (S), 205-214

Tester, R.F. & Morrison, W.R. (1990). Swelling and gelatinization of cereal starches. I. Effects of amylopectin, amylose and lipids. *Cereal Chemistry*, 67 (6), 551-557

Trater, A.M., Alavi, S. & Rizvi, S.S.H. (2005). Use of non-invasive X-ray microtomography for characterizing microstructure of extruded biopolymer foams. *Food Research International*, 38, 709-719

Ubbink, Job., Giardello, M.-I. & Limbach, H.-J. (2007). Sorption of water by bidisperse mixtures of carbohydrates in glassy and rubbery states. *Biomacromolecules*, 2007 (8), 2862-2873

Vallés Pàmies, B., Roudaut, C., Dacremont, C, Le Meste, M. & Mitchell, R. (2000). Understanding the texture of low moisture cereal products: Part I mechanical and sensory assessment of crispiness. *Journal of the Science of Food and Agriculture*, 80, 1679-1685

van Lengerich, B. & Larson, M.K. (2000). Cereal products with inulin and methods of preparation. U.S. Patent 6149965

Vergnes, B. & Villemaire, J.P. (1987). Rheological behaviour of low moisture molten maize starch. *Rheologica Acta*, 26, 570-576

Vergnes, B., Della Valle, G. & Tayeb, J. (1993). A specific slit die rheometer for extruded starchy products. Design, validation and application to maize starch. *Rheologica Acta*, 32, 465-476

Wang, S.M., Bouvier, J.M. & Gelus, M. (1990). Rheological behavior of wheat flour dough in twin-screw extrusion cooking. *International Journal of Food Science and Technology*, 25, 129-39

Wang, S.M., Casulli, J. & Bouvier, J.M. (1993). Effect of dough ingredients on apparent viscosity and properties of extrudates in twin-screw extrusion-cooking . *International Journal of Food Science and Technology*, 28, 465-479

Warburton, S.C. & Donald, A.M. (1992). Structure and mechanical properties of brittle starch foams. *Journal of Materials Science*, 27, 1469-1474

Weisser, H. (1986). Influence of temperature on sorption isotherms. In Maguer, M. & Jelen, P. *Transport Phenomena*. Elsevier Science, Ltd: London, PP. 662

Willett, J.L., Jasberg, B.K. & Wanson, C.L. (1995). Rheology of thermoplastics starch: effects of temperature, moisture concentration and additives on melt viscosity. *Polymer Engineering and Science*, 35, 202-210

Willett, J.L., Millard, M.M. & Jasberg, B.K. (1997). Extrusion of waxy maize starch: melt rheology and molecular weight degradation of amylopectin. *Polymer*, 38, 5983-5989

Williams, M, Landel, R. & Ferry, J. (1955). The temperature dependence of relaxation mechanisms in amorphous polymers and other glass-forming liquids. *Journal of the American Chemical Society*, 77, 3701-3707

Xanthos, M. (2005). *Modification of polymer mechanical and rheological properties with functional fillers*. In: Xanthos, M. (Ed.), Functional fillers for plastics. Wiley: Weinheim, pp. 17-38

Xie, F., Yu, L. Sing, B., Wnag, J., Liu, H. & Chen, L. (2009). Rheological properties of starches with different amylose/amylopectin ratios. *Journal of Cereal Science*, 49, 371-377

Yanniotis, S., Petraki, A. & Soumpasi, E. (2007). Effect of pectin and wheat fibers on quality attributes of extruded cornstarch. *Journal of Food Engineering*, 80, 594-599

Yoshii, H., Furuta, T., Noma, S. & Noda, T. (1990). Kinetic of Soy-protein denaturation by a temperature-programmed heat-denaturation technique. *Agricultural Biological Chemistry*, 54, 863-869.

Zhang, D. & Moore, W.R. (1997). Effect of wheat bran particle size on dough rheological properties. *Journal of the Science of Food and Agriculture*, 74, 490-496

Zeleznac, K.J. & Hoseney, R.C. (1987). The glass transition of starch. *Cereal Chemistry*, 64 (2), 121-124

Zobel, A. (1988). Molecules to granules: A comprehensive starch review. *Starch/Stärke*, 40 (2), S44-50

APPENDICES

FIGURES

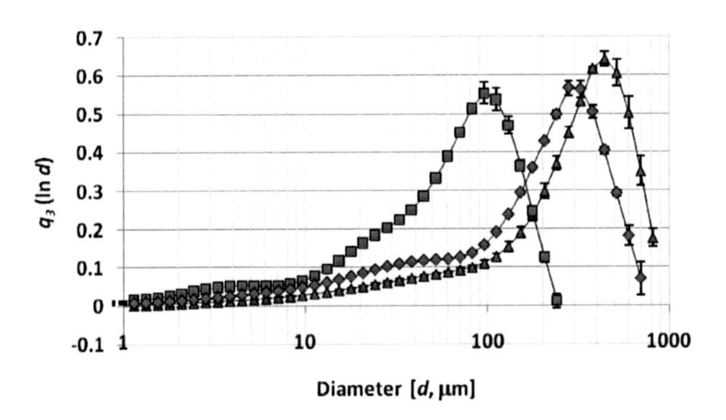

Fig. 10.1: Particle size distribution $\bar{q}_3 \ln(d)$ by volumes for the refined wheat flour (■), fine (▲) and coarse bran (♦) (results of triplicate measurement)

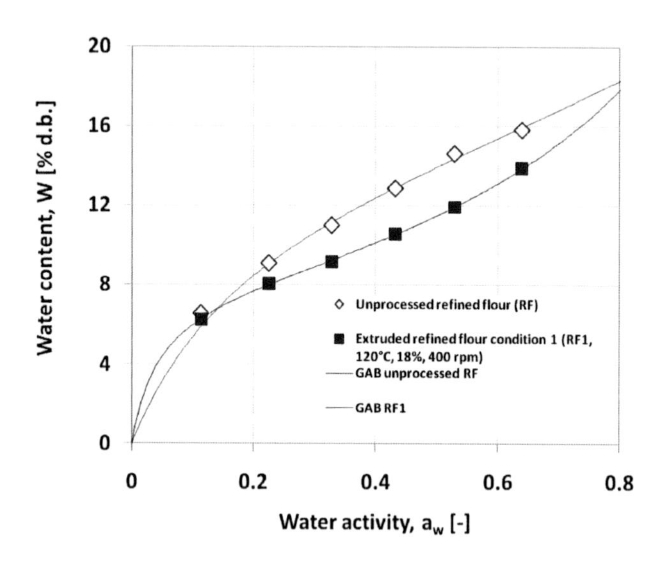

Fig. 10.2: Sorption isotherm representing the variation of water content W depending on the water activity a_w of unprocessed refined flour and refined four extruded at condition 1

Fig. 10.3: Extruded samples of refined flour (a), low bran concentration (b) and high bran concentration (c) under extrusion conditions 1 to 8 (from left to right)

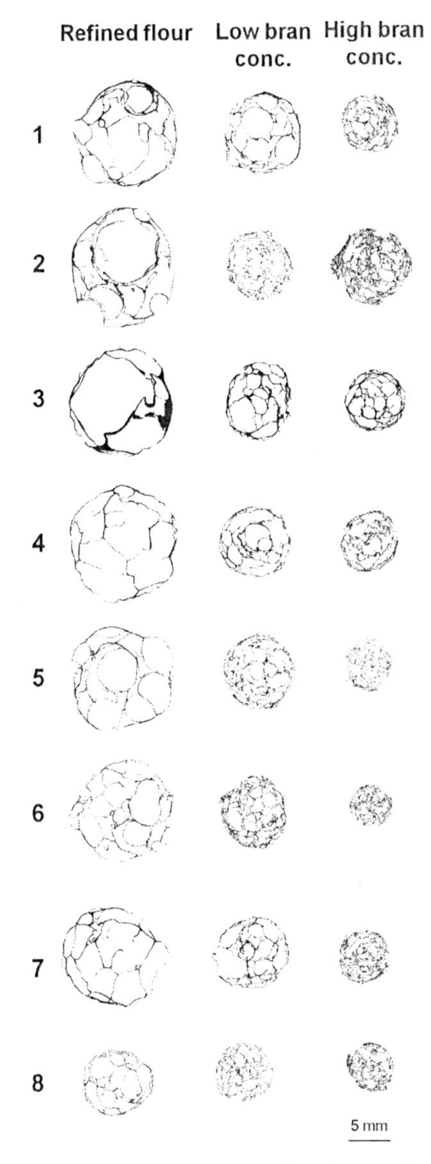

Fig. 10.4: **Cross sectional X-ray tomography pictures (6 μm thick) of extruded refined flour, low bran and high bran concentration samples depending on the extrusion conditions (conditions are shown in Chapter II in Tab. 3.2)**

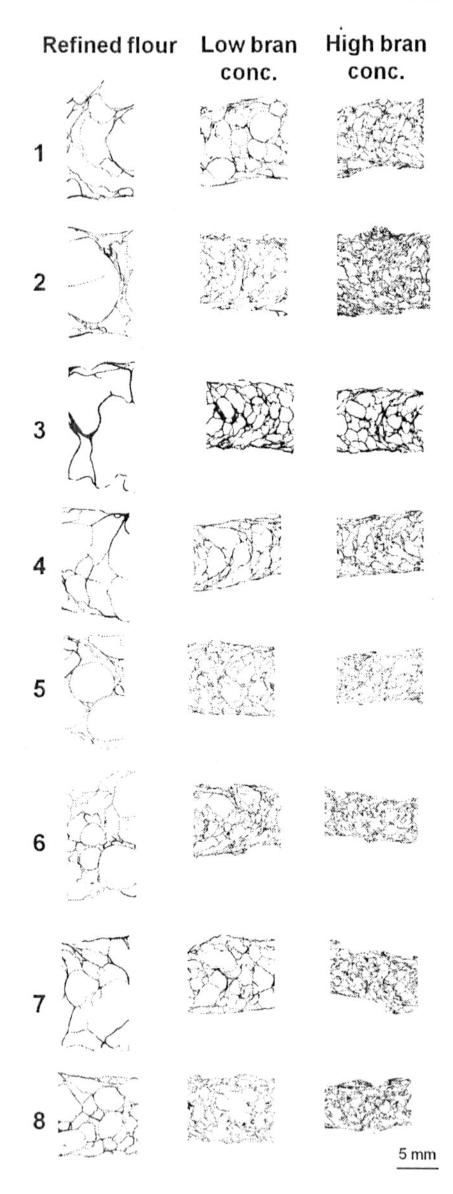

Fig. 10.5: Cross longitudinal X-ray tomography pictures (6 µm thick) of extruded refined flour (RF), low bran (LB) and high bran (HB) samples according to the extrusion conditions 1 to 8 (conditions are shown in Chapter II, Tab. 3.2)

TABLES

Tab. 10.1: Correlation factor r^2 between the experimental data and the experimental plan statistical model and least significant difference (LSD) for the refined flour (RF), low bran concentration (LB) and high bran concentration (HB) samples

		SME	WAI	WSI	VEI	SEI_{max}	SEI	LEI	SR	Log(MCS)	Log(MCWT)	Log(N_c)	Shear viscosity (30 s^{-1})	K	n	σ*	E*
R^2 [-]	RF	0.98	0.87	0.83	0.96	0.23	0.67	0.90	0.49	0.21	0.86	0.84	0.92	0.92	0.86		
	LB	0.98	0.92	0.91	0.96	0.61	0.57	0.90	0.88	0.73	0.87	0.92	0.90	0.73	0.67		
	HB	1.00	0.56	0.84	0.67	0.86	0.90	0.89	0.67	0.75	0.67	0.80	0.63	0.50	0.35		
	LSD*	7.4	0.9	3.7	2.5	6.1	3.4	0.51	17	0.50	0.20	0.84	405	15300	0.12	0.13	0.63

SME: specific mechanical energy, WSI: water solubility, WAI: water absorption index, VEI: volumetric, SEI_{max}: maximum sectional, SEI: sectional and LEI: longitudinal expansion index, SR: shrinkage ratio, MCS: mean cell size, MCWT: mean cell wall thickness, N_c: cell density, K: corrected consistency factor, n: power law index, σ^*: stress at rupture and E^*: elastic modulus (*units of LSD are the same as units of the parameter)

Tab. 10.2: Water solubility index (WSI), water absorption index (WAI), estimated water soluble starch (SWS) and rapid visco analysis (RVA) peak viscosity according to process conditions and bran concentration (refined flour (RF), low bran concentration (LB) and high bran concentration (HB))

	WSI [% d.b.]	WAI [% d.b.]	Estimated water soluble starch (SWS) [% d.b.]	RVA peak viscosity [cP]
RF1	34.2 ± 0.3	5.46 ± 0.02	43.6 ± 0.4	220 ± 20
RF2	41.5 ± 0.1	4.41 ± 0.02	52.9 ± 0.2	124 ± 4
RF3	31.8 ± 0.5	5.83 ± 0.10	40.5 ± 0.4	314 ± 28
RF4	35.6 ± 0.2	5.03 ± 0.02	45.3 ± 0.3	146 ± 4
RF5	33.0 ± 0.3	5.42 ± 0.08	42.0 ± 0.4	289 ± 16
RF6	32.9 ± 0.0	4.65 ± 0.00	42.0 ± 0.1	138 ± 6
RF7	24.3 ± 0.1	6.00 ± 0.01	30.9 ± 0.2	413 ± 11
RF8	21.3 ± 0.7	5.89 ± 0.10	27.1 ± 0.9	446 ± 1
LB1	30.6 ± 0.4	4.35 ± 0.04	44.1 ± 0.8	197 ± 11
LB2	36.9 ± 0.3	3.57 ± 0.05	53.2 ± 0.7	158 ± 14
LB3	26.7 ± 0.3	4.80 ± 0.06	38.5 ± 0.6	200 ± 3
LB4	30.8 ± 0.2	4.27 ± 0.09	44.4 ± 0.6	135 ± 6
LB5	29.3 ± 0.2	4.36 ± 0.06	42.2 ± 0.5	166 ± 14
LB6	35.0 ± 0.2	3.93 ± 0.03	50.4 ± 0.5	105 ± 6
LB7	22.3 ± 0.3	5.38 ± 0.02	32.1 ± 0.6	261 ± 4
LB8	23.7 ± 0.0	5.09 ± 0.01	34.2 ± 0.3	242 ± 7
HB1	21.8 ± 0.3	3.64 ± 0.09	40.2 ± 1.9	103 ± 9
HB2	33.3 ± 0.2	3.19 ± 0.02	61.4 ± 2.3	25 ± 8
HB3	13.1 ± 0.5	6.16 ± 0.13	24.1 ± 1.8	99 ± 6
HB4	27.5 ± 0.2	3.38 ± 0.03	50.7 ± 2.1	86 ± 3
HB5	24.2 ± 0.1	3.42 ± 0.03	44.6 ± 1.6	91 ± 5
HB6	34.2 ± 0.2	3.18 ± 0.2	63.1 ± 2.5	64 ± 2
HB7	23.7 ± 0.1	3.70 ± 0.01	43.8 ± 1.6	110 ± 2
HB8	25.1 ± 0.2	3.69 ± 0.00	46.3 ± 1.9	132 ± 11

Tab. 10.3: Guggenheim-Anderson-de Boer (GAB), Gordon & Taylor (G&T) and Flory-Huggins fitting parameters of raw materials according to their bran concentration (refined flour (RF), low bran concentration (LB) and high bran concentration (HB))

	GAB Model			G&T Model		Flory-Huggins Model
	K_{GAB} [-]	C_{GAB} [-]	W_m [%]	$T_{g,m}$ [°C]	k [-]	T^0_{pk} [°C]
Refined flour	0.39	13.4	14.6	194	5.3	255
Low bran conc.	*0.52*	*18.3*	*11.6*	*197*	*6.0*	*247*
High bran conc.	0.53	16.0	10.8	178	5.7	198
Bran	0.73	18.4	8.1	105	6.8	197

K_{GAB}: constant, C_{GAB}: constant, W_m: theoretical monolayer water content, $T_{g,m}$: glass transition of the dried material, k: Gordon-Taylor fitting parameters, : melting temperature at the peak of dried crystallites

Tab. 10.4: Guggenheim-Anderson-de Boer (GAB) and Gordon & Taylor (G&T) fitting parameters of extruded material at condition 1 and 8 (refined flour (RF), low bran concentration (LB) and high bran concentration (HB))

	GAB Model			G&T Model	
	K_{GAB} [-]	C_{GAB} [-]	W_m [%]	$T_{g,m}$ [°C]	$K_{G\&T}$ [-]
RF1	0.72	32.4	7.7	187	5.2
LB1	0.66	24.4	7.9	148	3.8
HB1	0.81	30.3	6.35	135	3.8
RF8	0.77	25.9	8.0	161	4.1
LB8	0.72	30.3	7.4	140	3.7
HB8	0.80	27.7	6.7	154	4.8

K_{GAB}: constant, C_{GAB}: constant, W_m: theoretical monolayer water content, $T_{g,m}$: glass transition of the dried material, k: Gordon-Taylor fitting parameters, : melting temperature at the peak of dried crystallites

Tab. 10.5: Melt pressure (*P*), specific mechanical energy (*SME*) and melt temperature (*T*) in the extruder when using the die (index *d*) or the rheometer (index *r*) at different process conditions and bran concentration (refined flour (RF), low bran concentration (LB) and high bran concentration (HB)) and corresponding melt flow power law index (*n*), corrected consistency factor (*K*) and shear viscosity at 30 s^{-1} (η) of the extruded melt

	P_d [MPa]	P_r [MPa]	SME_d [kJ kg^{-1}]	SME_r [kJ kg^{-1}]	T_d [°C]	T_r [°C]	n [-]	K [10^3 Pa s]	η (30 s^{-1}) [10^3 Pa s]
RF1	14.8 ± 0.3	14.8 ± 0.3	506 ± 5	560 ± 0	130.7 ± 2.5	149.4 ± 2.8	0.10	46.0	2.15
RF2	8.4 ± 0.1	8.4 ± 0.1	607 ± 11	678 ± 2	138.1 ± 0.4	158.8 ± 0.5	0.08	33.0	1.45
RF3	11.4 ± 0.1	11.4 ± 0.1	371 ± 9	472 ± 2	118.0 ± 0.6	134.4 ± 0.1	0.12	29.2	1.46
RF4	6.8 ± 0.1	6.8 ± 0.1	449 ± 6	544 ± 2	132.7 ± 1.2	147.5 ± 0.1	0.12	23.0	1.15
RF5	10.4 ± 0.1	10.4 ± 0.1	319 ± 3	448 ± 2	170.7 ± 0.7	182.1 ± 0.5	0.21	14.7	1.00
RF6	6.1 ± 0.2	6.1 ± 0.2	454 ± 9	521 ± 2	171.0 ± 1.5	178.7 ± 0.6	0.27	10.1	0.84
RF7	7.8 ± 0.1	7.8 ± 0.1	233 ± 5	319 ± 2	163.5 ± 0.8	175.7 ± 1.1	0.26	8.5	0.67
RF8	5.6 ± 0.1	5.6 ± 0.1	280 ± 6	438 ± 5	170.7 ± 0.2	182.1 ± 0.8	0.19	9.5	0.61
LB1	13.5 ± 0.1	13.5 ± 0.1	559 ± 6	540 ± 0	130.8 ± 0.2	151.3 ± 0.3	0.09	52.3	2.33
LB2	7.8 ± 0.2	7.8 ± 0.2	630 ± 18	622 ± 1	143.5 ± 0.2	158.6 ± 0.2	0.11	26.1	1.26
LB3	10.3 ± 0.1	10.3 ± 0.1	430 ± 5	415 ± 2	124.6 ± 2.1	133.5 ± 0.7	0.18	24.4	1.50
LB4	6.5 ± 0.1	6.5 ± 0.1	511 ± 7	520 ± 2	133.8 ± 0.6	149.0 ± 0.9	0.09	24.5	1.11
LB5	10.2 ± 0.1	10.2 ± 0.1	394 ± 4	391 ± 2	179.8 ± 0.7	184.9 ± 0.2	0.27	17.0	1.40
LB6	5.8 ± 0.2	5.8 ± 0.2	516 ± 9	515 ± 2	168.7 ± 2.5	180.9 ± 0.2	0.23	10.9	0.81
LB7	7.1 ± 0.2	7.1 ± 0.2	271 ± 3	258 ± 1	163.1 ± 0.4	167.2 ± 0.5	0.23	14.9	1.07
LB8	4.9 ± 0.1	4.9 ± 0.1	406 ± 5	414 ± 1	164.8 ± 2.8	179.8 ± 0.6	0.12	13.3	0.66
HB1	13.3 ± 0.2	14.1 ± 0.1	587 ± 6	570 ± 2	135.9 ± 0.5	149.4 ± 0.3	0.13	50.5	2.60
HB2	7.8 ± 0.1	8.2 ± 0.0	706 ± 36	657 ± 3	147.6 ± 0.4	171.9 ± 3.7	0.12	21.9	1.11
HB3	10.1 ± 0.2	11.1 ± 0.1	435 ± 6	475 ± 1	126.6 ± 1.7	139.3 ± 1.1	0.08	53.4	2.35
HB4	6.5 ± 0.1	6.4 ± 0.1	555 ± 6	511 ± 3	136.1 ± 0.4	150.5 ± 0.0	0.24	13.9	1.05
HB5	9.0 ± 0.2	9.9 ± 0.1	412 ± 7	436 ± 1	175.9 ± 0.6	182.6 ± 1.0	0.10	35.3	1.66
HB6	6.0 ± 0.1	6.3 ± 0.1	567 ± 35	547 ± 3	170.8 ± 0.8	174.5 ± 1.3	0.05	31.8	1.26
HB7	7.3 ± 0.1	7.4 ± 0.0	292 ± 5	273 ± 1	165.7 ± 0.9	173.4 ± 1.2	0.18	14.9	0.90
HB8	5.4 ± 0.1	7.2 ± 0.0	415 ± 7	496 ± 2	171.1 ± 0.8	174.9 ± 0.0	0.13	23.1	1.18

Tab. 10.6: Bulk expansion properties (maximum sectional expansion index (SEI_{max}), sectional expansion index (SEI), shrinkage ratio (SR), longitudinal volumetric expansion index (VEI) and relative density (*D*)), cellular structure characteristics (mean cell wall thickness (MCWT), cell density (N_c) and mean cell size (MCS)) and foams mechanical properties (stress at rupture σ^* and elastic modulus E^*) at different process conditions and bran concentration (refined flour (RF), low bran concentration (LB) and high bran concentration (HB))

	Bulk expansion properties						X-Ray Tomography			Mechanical properties	
	SEI_{max} [-]	SEI [-]	SR [%]	LEI [-]	VEI [-]	*D* [-]	MCWT [μm]	N_c [cm^{-3}]	MCS [μm]	σ^* [MPa]	E^* [MPa]
RF1	25.2 ± 2.8	22.5 ± 1.9	11	0.62 ± 0.03	13.9 ± 1.7	0.066 ± 0.008	70	100	1610	0.58 ± 0.13	15.9 ± 4.9
RF2	29.1 ± 4.7	23.5 ± 0.9	19	1.00 ± 0.04	23.4 ± 1.3	0.038 ± 0.002	45	430	3650	0.27 ± 0.08	5.3 ± 1.9
RF3	30.0 ± 4.4	17.7 ± 0.9	41	0.35 ± 0.02	6.2 ± 0.4	0.150 ± 0.010	265	2	3220	0.82 ± 0.20	56.5 ± 13.8
RF4	33.3 ± 3.3	25.0 ± 1.1	25	0.80 ± 0.05	20.0 ± 0.7	0.044 ± 0.001	85	30	1900	0.37 ± 0.08	11.0 ± 4.2
RF5	31.4 ± 3.8	21.7 ± 1.1	31	0.95 ± 0.06	20.7 ± 1.5	0.043 ± 0.003	45	320	2180	0.32 ± 0.07	9.5 ± 3.0
RF6	27.3 ± 4.6	15.7 ± 1.0	42	1.89 ± 0.11	29.8 ± 2.7	0.029 ± 0.003	30	620	1850	0.19 ± 0.04	9.1 ± 2.6
RF7	28.6 ± 4.8	13.5 ± 0.9	53	0.86 ± 0.04	11.5 ± 1.1	0.076 ± 0.007	75	50	1730	0.67 ± 0.14	18.5 ± 5.8
RF8	13.4 ± 2.0	10.1 ± 0.4	25	2.00 ± 0.10	20.2 ± 1.2	0.041 ± 0.002	30	740	1940	0.86 ± 0.19	19.7 ± 7.1
LB1	11.2 ± 0.5	11.6 ± 0.4	0	0.88 ± 0.05	10.2 ± 0.3	0.089 ± 0.002	50	510	1140	0.87 ± 0.15	40.8 ± 7.4
LB2	9.9 ± 0.6	8.7 ± 0.2	12	1.55 ± 0.06	13.6 ± 0.5	0.066 ± 0.002	30	800	540	0.47 ± 0.09	20.0 ± 4.6
LB3	11.9 ± 0.7	7.7 ± 0.2	35	0.64 ± 0.02	4.9 ± 0.2	0.183 ± 0.007	110	120	820	1.80 ± 0.37	105.6 ± 22.7
LB4	13.2 ± 0.7	10.3 ± 0.4	22	0.99 ± 0.06	10.2 ± 0.4	0.086 ± 0.003	55	220	890	0.99 ± 0.15	41.6 ± 11.3
LB5	9.9 ± 0.7	9.4 ± 0.4	5	1.57 ± 0.10	14.8 ± 0.9	0.061 ± 0.004	30	1250	590	0.71 ± 0.16	28.8 ± 7.1
LB6	8.7 ± 0.6	8.2 ± 0.3	6	2.57 ± 0.13	21.0 ± 0.7	0.042 ± 0.001	25	3120	470	0.44 ± 0.10	18.4 ± 4.2
LB7	12.5 ± 0.8	8.4 ± 0.5	32	1.18 ± 0.09	9.9 ± 1.0	0.089 ± 0.010	40	380	850	1.33 ± 0.31	57.1 ± 14.4
LB8	8.0 ± 0.3	4.8 ± 0.2	39	2.67 ± 0.08	12.9 ± 0.9	0.065 ± 0.005	25	2120	410	1.15 ± 0.18	45.6 ± 7.6
HB1	7.0 ± 0.3	5.9 ± 0.3	15	1.28 ± 0.03	7.6 ± 0.4	0.121 ± 0.006	40	1610	380	1.00 ± 0.12	67.4 ± 10.8
HB2	4.5 ± 0.3	3.9 ± 0.3	13	2.66 ± 0.18	10.5 ± 0.9	0.084 ± 0.007	30	3650	210	0.55 ± 0.17	26.9 ± 8.2
HB3	7.1 ± 0.3	4.9 ± 0.3	31	0.91 ± 0.02	4.5 ± 0.3	0.202 ± 0.014	90	170	580	2.93 ± 0.45	224.9 ± 46.4
HB4	6.9 ± 0.3	4.9 ± 0.3	29	1.51 ± 0.07	7.4 ± 0.6	0.119 ± 0.009	40	860	400	1.33 ± 0.15	81.7 ± 14.4
HB5	5.0 ± 0.3	3.3 ± 0.5	35	2.43 ± 0.36	7.9 ± 1.1	0.114 ± 0.019	25	1260	270	0.68 ± 0.16	41.8 ± 11.4
HB6	4.1 ± 0.2	1.6 ± 0.2	62	5.42 ± 0.69	8.5 ± 0.9	0.105 ± 0.011	25	4980	230	0.50 ± 0.14	22.7 ± 7.1
HB7	5.9 ± 0.3	4.1 ± 0.3	31	1.80 ± 0.06	7.4 ± 0.6	0.117 ± 0.010	35	1710	280	1.48 ± 0.26	82.7 ± 18.7
HB8	4.3 ± 0.3	2.2 ± 0.2	49	3.48 ± 0.08	7.7 ± 0.8	0.111 ± 0.011	30	2830	230	1.10 ± 0.27	54.9 ± 18.0